AN INTRODUCTION TO
PROCESS CONTROL
AND DIGITAL
MINICOMPUTERS

PETER L. GINN

AN INTRODUCTION TO
PROCESS CONTROL
AND DIGITAL
MINICOMPUTERS

gpgc Gulf Publishing Company
Book Division
Houston, London, Paris, Tokyo

To my own personal editor, Ann Clotilde,
without whose drawings and typing
this book would not be.

An Introduction to

PROCESS CONTROL AND DIGITAL MINI COMPUTERS

Library of Congress Cataloging in Publication Data

Ginn, Peter L.
An introduction to process control and digital minicomputers.

Includes index.
1. Chemical process control—Data processing.
2. Minicomputers.   I. Title.   II. Title: Process control and digital minicomputers.
TP155.75.G56   660.2'81   82-6036
ISBN 0-87201-180-1   AACR2

CONTENTS

PREFACE

This book's major purpose is to familiarize chemical production engineers and managers of chemical process plants with the unfamiliar task of controlling a chemical process by means of a digital computer. No previous knowledge of digital computers is assumed. Two chapters are included, intended for chemical engineers, which explain the terminology, the architecture, and the overall operation of computer systems. Another purpose of the book is to help electrical and electronics engineers, and computer programmers likewise involved with such tasks, to familiarize themselves with chemical process plants. For non-chemical engineers, two chapters are included which cover simple, conventional instrumentation, the elements of flow control, and the description of a chemical process.

There is no *best* way to construct, operate, and control a particular chemical process. Consequently, the methods presented in this book are not to be held as definitive, but are offered as a stimulus to practitioners, as guidelines to novices, and as representative of one complete exercise to the uninitiated.

The process described is the "Witches' Brew Process," a process that not too many people are aware exists. As no reader has preconceived notions about how this Transylvanian process should be designed or controlled, all readers should be able to concentrate on the implementation of digital computer process control rather than the process itself.

Peter L. Ginn
April 1982

INTRODUCTION

You are reading this book because you are interested in automating a chemical process with a programmable processor. Perhaps you are a chemical engineer who already has an approved computerization project scheduled, or installed, and you are casting around as how best to proceed. You are not sure what resources it will demand, or how to check on progress; perhaps you feel imaginative solutions might be feasible if only you knew how. On the other hand, you may have no experience in this field of endeavor and no immediate plans but are curious about chemical process automation with programmable processors. Perhaps most of this technology was invented after you got your degree and you find the whole thing, with its cryptic terminology, very, very confusing. Or perhaps you are a plant manager with a bottlenecked plant or a plant with utility consumption, by your reckoning, excessive. You know that if you could hold that stoichiometric ratio steady, shift after shift, you could set records, but your conventional instrumentation just cannot deliver.

Whether you are a manager who wants just a brief exposure to the material, a person looking for ideas, or an engineer who needs a good grounding in the subject, there is material in this book for you. There are several ways to read it.

You may read it lightly, skipping any of the more difficult material. This will give you, if you are a manager, for example, a feel for the endeavor and some of its vocabulary. Or you may read it thoroughly. This will give a chemical engineer about to start his own project enough hindsight to make it successful. Or you may read it frequently to understand all of the material within the book, and then use that knowledge. Using that knowledge will turn your machines into bus bangers.

At the end of the book, the novice should have answered most of his questions, and the expert should have had some of his assumptions challenged. For both, the book contains material which should improve their results. In it is described the installation of a computer in a process—from proposal to total implementation.

This keeps the book practical, it allows for a better explanation of the alternatives, and it enables some of the ramifications of the actions taken to be explored. I immediately acknowledge that the way the project is handled in this book is not the only way the project could have been handled. But the approach I describe is easily followed, and it does produce results.

Most chemical engineers are acquainted with one or another attempt at computer control of a chemical process. Their overall perception is that it is extraordinarily difficult to get meaningful results—particularly results that will not deteriorate over time. Their perception is correct. But the prospective user can avoid many of the difficulties if at the beginning the project is approached correctly, and if he has a grasp of all the intricacies involved. It is hoped that the book will help him in both ways.

Always think of a project as consisting of four aspects, and view them as you would the four faces of a tetrahedron:
1. The electronic digital computer (the pieces of hardware themselves).
2. The operating system (the manufacturer-supplied software that makes the computer run).
3. The application software (written in the programming language of the user's choice).
4. The actual application (the use to which the other three are to be put).

Each bears on the other three; any one touches the other three; none can be chosen in a vacuum. Please bear this concept in mind throughout this book.

No specific manufacturer's machine will be detailed nor will any particular language be emphasized. Those readers who may be disheartened at this point may rest assured that the underlying principles are understood more easily and have a more universal application than any machine or language specifics. History has also been avoided. Historical circumstances and commercial convenience have obfuscated the "how" of what is a perfectly straightforward machine. If the software for the machine is considered as one giant algebraic expression and if the machine itself is considered as an automatic evaluator of that expression, then the interdependence of the two can be recognized. Furthermore, it should be realized that the expression is not the evaluation.

Three features of electronic digital computers must be mentioned so that all readers are aware of them. First, the computer will follow the program instructions in sequence automatically. Second, the program's instructions must be coded in numerical form for the machine to be able to use them. Finally, the machine does not handle numbers directly—it only handles registers, and it is these registers that contain the numbers. The elegant simplicity of these three features is what gives digital computers their power and universality. The simplicity is undiminished whatever the size or sophistication of the computer.

It is easy enough to recognize in a chemical process how a human-supervisory/human-controlled system works and what information it needs. Chemical engineers have at hand notebooks, handbooks, hand-held calculators, previously written algorithms and data tables, wall clocks, paper and pencil, appointment calendars, operating manuals, logbooks, foremen, and operators. With a knowledge of process tank gauges, material flows, pressures, and temperatures, the engineer can calculate optimum conditions for the plant equipment and the process. He can coordinate the production at the specified rate and initiate corrective action whenever deviations occur. This is done relatively slowly, but very visibly. The whole system is understood by all the people in contact with it.

In a computer-supervisory/computer-controlled system the workings and the needs are much more difficult to recognize. Indeed, there is a whole new terminology to learn. Such a system requires hardware and software. The hardware consists of the CPU, the memory, the hard disks, the floppy disks, the clocks, the communication interfaces, the video generators, the keyboards, the printers, the television monitors, the plant analog input interface device, the computer digital output interface device, and the paper-tape reader punch. The software consists of the system director program, the loading program, the start program, the patching program, the editing program, the math program, the input conversion program, the control program, the calculation program, the alarm-raising program, the CRT driver program, the teletypewriter driver program, the report-generating program, the data-gathering program, the plotting program, and a whole handful of datasets for data storage and display information purposes. Hidden in this software are the control algorithms that must be developed, the provisions that must be made to accommodate changing targets and limits, the handling of requests for displays and reports, and the flexibility needed as perceptions (as to how the process should be run) are modified. The flow of information from field measurement into the computer, and its storage and eventual regeneration back out of the computer onto paper or into video, is both fast and unseen. Few people in contact with the system understand its workings.

This book addresses these problems in the following ways:

Chapter 1 gives reasons as to why installing a computer is worth the effort required. It explores the software dilemma. It lists 16 questions to which answers must be found before any computerization project is initiated. A set of rules is given for handling the project. The chapter closes with some comments about programming productivity and program productivity.

Chapter 2 looks briefly at process instrumentation. Some chemical engineering fundamentals such as fluid flow measurement and fluid flow regulation are reviewed. The control of flow in pipes is studied at length. The conventional PID algorithm is examined.

Chapter 3 describes the Witches' Brew Process, the process that will be computerized in the course of this book. There is a process flow and instrument diagram. Inventories of equipment and instruments are included. The process's reaction and purification steps are described in detail. Balances on material and energy are given. Finally, the capital costs and process economics are included.

Chapter 4 looks at the hardware: memory, bulk storage, video display terminals, keyboards, paper-copy terminals, process input, process output, and information transmission. A system is configured.

Chapter 5 elucidates computer concepts and describes simple programming, floating-point math, and typical system software. A software inventory for the Witches' Brew Process is given. The iterative process of software generation is described thoroughly.

Chapter 6 returns to the Witches' Brew. The objectives of the project are outlined. The proposed computer system is inventoried and described, and the shopping lists are prepared. The chapter concludes with the economics of the proposal.

Chapter 7 examines the computerized Witches' Brew Process from process measurements through input, the mathematics, and output, to expected results. Several ways are suggested to generate data logging, conduct statistical studies, and plot optima.

Chapter 8 closes the control loops. It compares PID and non-PID algorithms, particularly decision trees. The deficiencies of analog controllers and the power of programmable processors are described. It explains what is involved in hierarchical control and total process control. The chapter ends with a comparison of diminishing returns, marginal returns, and most-effective returns.

Chapter 9 closes out the book with comments on yesterday and tomorrow. What lessons can we learn from the past? Where is the dizzying whirl of microelectronics taking us? Is there a nascent profession emerging? What software is available?

There are several underlying ideas presented in this book of which the reader should be aware.

First, the computer is a process-research aide of unimaginable skill. With a suitably designed data base, the graphing of two variables, the trending of a measurement, or the statistical analysis of a group of data are easily implemented. Any knowledge generated in this way gives an illuminating insight to the process. Using high-resolution plotters and theoretically sound statistical tests, it is no longer necessary to rely on conventional instrument paper charts or instinct to distinguish the good from the bad. This means that the understanding, the progress, the value of the control strategies, and the techniques being implemented can be proven and documented. A process upset can be replayed in slow motion or stop frame. The process's inner happenings can be picked apart.

Second, making the digital computer a substitute for an analog controller with PID algorithm is not taking advantage of the machine's fullest capability. A correctly conceived computer-based process control effort will use the human operator's actions as a guide, not the workings of an analog-controller. As a consequence, there is no grand algorithm which can be universally implemented. Instead, what can be implemented are imaginative and sophisticated decision trees of unlimited variation that will handle both static and dynamic changes. Decisions within the tree are based on whether the deviation is correctable and worth correcting. How and where a valve must be positioned to reduce a deviation, to maximize an opportunity, or to improve the process are all considered in this book. It must be realized that a computer can handle not only many inputs simultaneously, but also several errors simultaneously. And, even more powerfully, the computer can handle process dead-time. That is worth repeating: *the computer can handle process dead-time.*

Third, the control valve is a variable restricting-orifice. It can be manipulated in any way that one sees fit. It can be slammed shut. It can be thrown wide open. It can be opened very, very slowly. It can be closed just a hair. It does not have to respond to every deviation. It can be positioned pre-emptively. It is a valid control strategy to hold the valve position rock-steady. It is an equally valid control strategy to jiggle the valve back and forth over a limited range.

Fourth, the effort of implementing computer control is subject to severe diminishing returns. It may take only 10% of the programming effort to give improved process control for 90% of the time. It may take nine times the original effort to make the computer give better process control than the next best alternative for the remaining 10% of the time. Computer control also produces great rewards. Perhaps raw material savings will pay for the installation and programming effort in five months. Perhaps a batch reactor's productivity will be doubled.

Fifth, computer operation and process control are very difficult to describe in the general case. But to refer to any one particular system makes for specifics that are of limited value. Consequently, this book describes an imaginary process and a virtual processor. Neither exists, in fact, but in general terms they are representative of a chemical process and a digital computer.

Sixth, there is a lot of confusion and ignorance concerning digital computers and process control. Some confuse "computer-the-CPU-chip" (which is impotent alone) with "computer-the-system" (a combination of software, hardware, and peripherals as powerful as your imagination will allow it to be). Some confuse "process control theory" (Nyquist diagrams, Laplace transforms, and PID algorithms) with "process control fact" (an experienced operator's ability to line-out an upset plant). No single book can address itself to helping everyone out of his tar-pit, but it is hoped that this book will be a *first* step for some toward explaining the terms used, describing the operation of the equipment, and getting started in this field of endeavor.

At the end of this book you may not be able to tell a UART from an FPP, a T-state from a disk revolution. You still will not appreciate the merits of RATFOR versus FORTRAN IVH, or be able to distinguish a compiled BASIC from an integer BASIC—but you will be able to close a loop, understand the difference between DDC and SSC, and be able to configure a system that will meet *your* needs.

Last, flow control cannot be dismissed as simple. Temperature control, level control, and pressure control inevitably involve the control of fluid flow. Good flow control is THE fundamental building block to good process control.

NOTE: This book distinguishes between a computer cycle and a process period.

A computer cycle is the periodicity of the computer's actions. The computer scans the chemical process, it calculates, it makes logical decisions, it makes adjustments, it updates various displays. It is then quiescent until this sequence of actions is repeated. That cycle of events is a computer cycle. It is measured in seconds.

A process period for a chemical process is that period of time that must elapse before any distinguishable change in a flow FROM the process is recognized following a change in some flow TO the process. It can also be taken to mean the time needed for any chemical or physical process to reach a stable equilibrium following a deliberate change of conditions. It is measured in minutes.

The slow progress and the disappointing results of computer control were mentioned earlier. Several reasons can be given:

1. Oversimplification of the Problem

The digital control of a chemical process involves much more than the digitization of analog controllers. It requires a thorough examination and analysis of the process first. Not being able to calculate to six figures of significance does not limit the performance of analog controllers. Possibly the plant itself limits their performance.

2. Overexaggeration of the Prowess of Computers

They are fast, they do not become bored, they are accurate to the limit of their means; but they are stupid, cannot think for themselves, and they must have everything letter-perfect.

3. Misplaced Emphasis on the Latest Hardware

All equipment requires a learning curve. It is better to be high up on the learning curve of an "old" piece of electronic equipment that is reliable than it is to be repeatedly on the low portions of curves of "new" and unreliable equipment. The human instinct to have the latest and the best is self-defeating.

4. The Detraction of Supersophisticated Software

The temptation to race ahead and develop elegant solutions to resource sharing and problems of allocation, impressive displays of blinking colors and pulsing lines, and massive comprehensive management information systems—all this has been at the expense of understanding and strengthening the most basic assumptions about the processes and their control. Perhaps the foundations are wrong.

5. Improper Resource Allocation

By far the biggest expense in any computerization project is lost opportunity cost: the dollar savings that are not being generated because the project is not complete. The second greatest expense is software development. The third greatest expense is the *process* interface. One of the smallest identifiable expenses is the CPU and its memory. Do not haggle for the cheapest CRT or cheapest CPU. Attention should be addressed to getting superfast equipment delivery, to identifying goals in software (a simple and quick approach), and finally to developing and installing better process interfaces.

6. Use of the Wrong People

The common myth is that people with people-problems can work with machines where there is little people-involvement. They can become productive employees and yet not upset morale. *This is a disastrous error.* People implementing process automation by programmable processor should be superior to everyone else at human relations. They must persuade, educate, and train nearly everyone else who comes into contact with the machine. This includes craftsmen, instrument men, operators, engineers, managers, owners, and in some circumstances even vendors.

7. Historical Circumstances

Engineers approach industrial chemistry after solving involved problems on large main-frame computers in the FORTRAN language using batch mode, and with programs and data entered on cards. Chemists approach industrial chemistry after organizing and supervising laboratory experiments with minimally configured mini-computers. The programs were entered in BASIC under time sharing, and data was read-in in real time. Neither approach is good preparation for process computers. The differences between the environments of a chemist in a lab, a person working on computer control of a process, and an engineer in a design office are listed in Tables I-1, I-2, and I-3. Each approaches the process computer

Table I-1
Approach: Chemist/Laboratory/Monitoring of Experiments/University

Purpose: To generate facts
Time Budget: Not critical, not time-limited
Dollar Budget: Severely restricted

Installation: Small data storage requirements (size of one experiment)
: Can halt machine at any time at user's convenience
: Non-automatic program loading acceptable
: Cold starts inconvenient, but not a disaster
: Small communication interfaces (2 chan @ 110 baud)
: Small "front-end" (16 analog input; 16 DI/DO)
: Consequences of machine failure not severe
: Real-time system has some flexibility; batch possible
: Many applications of configuration
: Many applications of software, could be used elsewhere
: Math packages are involved, somewhat slow, and of great precision
: Machine's output itself is required result
: Affords people productivity (fewer required)

Environment: Infrequent people-turnover
: User very knowledgeable about computer system
: Unrestricted flows of user knowledge
: Labor intensive
: Very few users
: Non-team orientation

Overall: Cannot afford computer hardware; has time to develop software; chooses own hardware

Table I-2
Approach: ?/Manufacturing/Control of Process Operation/Industry

Purpose: To produce chemicals
Time Budget: Always insufficient
Dollar Budget: Elastic; capital available for good projects

Installation: Large on-line data storage requirements (many long histories)
: Machine must be "up" at all times
: Must have fully automatic program loading
: Ill afford system regeneration, too much history lost
: Medium communication interface (8 chan @ 2400 baud)
: Large "front-end" (320 analog input; 144 DI/DO)
: Machine failure very serious
: Very demanding real-time only
: Individual configuration unique
: Custom software, not usable elsewhere
: Math packages are simple, fast, and relatively accurate
: Required result comes from machine's output affecting real world
: Affords raw material productivity (more chemical)

xv

Environment: Frequent people-turnover
: Users cannot afford to be ignorant of system
: User knowledge very restricted distribution, not easily passed on
: Capital intensive, people shy
: Many users, some trained, not all competent
: Orientation is to large teams

Overall: Can afford hardware; no time to develop software; chooses own hardware

Table I-3
Approach: Chemical Engineer/Office/Calculations/Commerce

Purpose: To design equipment
Time Budget: Comfortable
Dollar Budget: Some constraint

Installation: Large off-line data storage requirements
: Machine can go down, some inconvenience to users
: Manual loading of programs from cards & tape acceptable
: System regeneration—minor inconvenience
: Large communication interface (32 chan @ 300 baud)
: No "front-end"
: Machine failure, an interruption only
: Time-sharing (of a forgiving nature) and batch
: Few of like configuration
: Packaged software; software re-usable elsewhere
: Math packages complex and of sufficient accuracy
: Machine's output is required result
: Affords people productivity (more & better calculations)

Environment: Some people-turnover
: Users totally ignorant of computer system
: User knowledge has restricted distribution, but easily passed on
: Labor intensive (but not considered as such)
: Many users, all trained, most competent
: Orientation is to small teams

Overall: Hardware is not a consideration; software developed as necessary

differently. The engineer concentrates on sophisticated programs with little appreciation of the problems of real-time programming. He is tempted to buy packaged solutions to save time, and he has a predilection toward overly large, powerful systems. The chemist concentrates on undersized, inexpensive machines with straightforward programming and self-development of solutions.

With hindsight, it is easy to say that the best approach is down the middle, neither to one side nor the other. Avoid the system that is involved. It will be too difficult to get off the ground. Avoid the system that is too small. It will be outgrown quickly. What exemplifies the middle approach? That is still under active discussion by practitioners.

8. Ingredients for Success

Although the computers themselves are mass-produced and inexpensive, inculcation of how to use them is still individualized and expensive. It is a practical learning experience (particularly so as the technical literature and documentation sometimes includes mistakes), is not usually self-explanatory, and does not always contain the greatest clarity of thought.

One cannot succeed without a high productivity in generating application software. High productivity comes from:

1. An operating knowledge of the peripherals (CRT, printer) as installed
2. A knowledge of the computer's "operating system"
3. A knowledge of the machine's EDIT and BATCH facilities
4. Fluency in the chosen language
5. Typing skills
6. The capability to flow chart logic diagrams
7. An understanding of the organization of the processor
8. An understanding of the communication protocols in use
9. And a machine with good *UP*-time

One cannot succeed without getting results from the application software. Good results stem from:

1. Good applications software
2. Knowing and understanding the process processor and *all* its peripherals
3. Knowing and understanding the process processor's "operating system"
4. Knowing and understanding the chemical process and plant
5. Knowledge, understanding, and creativity in process control

1
THE DECISION
TO COMPUTERIZE

Why should anyone want to automate a chemical process with a programmable processor? Answer: ECONOMICS! The economics are very attractive. But the wary reader will want to explore this further. What resources must he commit? What are the returns? How big will the system be? How powerful is it? What exactly will he be buying? Why is the trade press not full of success stories?

What Are the Returns?

The quality of the product will be improved; the production capacity of the plant will be increased by 5%. The utility consumption per unit of product will be decreased by 10%. The raw material losses will be reduced by 20%.

The bold, round numbers are representative of the public numbers available in articles written about successful projects[1-3] and the more private numbers that surface at computer control courses and conferences. Individual projects may have figures that differ ± 100% from those given and will have different weightings of the individual contributions, but overall the percents already presented will stand closer examination. They do testify to the benefits that can be expected when a digital minicomputer and a process are combined. It can happen that some plants (especially discontinuous batch reactors) will show incredible productivity improvements in capacity, and other plants will register hardly any improvement at all. The results are very much a function of the initial conditions within the plant, perhaps more a function of them than a function of the power of the computer installed.

Overall installation of a computer control scheme leads to a lessened consumption of utilities, a greater utilization of the raw materials (because of

1

the reduced raw material losses), and a concomitant overall increase in the total throughput of the plant.

Can the source of the savings be pinpointed? No, it is an accumulation of many minor contributions: a fraction of a percent here, a fraction of a percent there . . .

Better stoichiometry
Improved controllability
Closer brinkmanship on specifications
Reduced reflux ratios
Greater process stability
Less thermal cycling of the equipment
Less spoiled material
Higher distillation tray efficiencies
Overall higher levels of total plant coordination
Higher furnace efficiencies
Better contingency preparedness
Faster startups
Real-time economic optimization
Better appreciation of process-limiting parameters
The uncovering of unsuspected bottlenecks
Marginal equipment run consistently within its performance limits
Less product giveaway
Production rate changes made in better balance and faster

What Are the Initial Costs?

If it is assumed that the computer installation will direct the operation of 24 control valves and will monitor 320 plant variables (flows, temperatures, and pressures), the approximate cost for the Witches' Brew Process (see Chapter 3) will be about $397,000. For that sum the plant will be totally self-supporting in computer services. The plant will have made its instrumentation computer-compatible. It will have prepaid 12 man-months of onsite software modification (see Table 1-1).

What Are the Hardware Maintenance Costs?

About 1% per month of the value of the hardware, or about $16,000 per year in this instance. This cost pays for both parts *and* labor.

Table 1-1
Costs of Installation ($)

Making plant instrumentation computer-compatible		160,500
Computer #1 processor	30,000	
Computer #2 processor	30,000	
Computer #1 mass storage	13,000	
Computer #2 mass storage	13,000	
Operator/engineer interface computer #1	31,000	
Operator/engineer interface computer #2	17,000	
Process plant interface computer #1	20,600	
Process plant interface computer #2	4,520	
Total #1	94,600	
Total #2	64,520	
	Total hardware	159,120
Sundry manuals, spares, cabinets, cables, power supplies: included		
Software, non-installation specific:	21,000	
Software, installation specific: initial	31,500	
Software, installation specific: modifications	21,000	
Associated fees	3,978	
	Total software	77,478
	Grand Total	397,098

How Much Manpower Must I Commit to the Project?

At least three man-months to change the instrumentation and to connect field measurements to the computer system. You will also need one full-time person to be responsible for installation, commissioning, and development of the complete system. The system will continue to need one full-time person not only to exploit the potential of the system, but to exploit fully the broadened capability of the chemical process. These people are the essence of the project and should be chosen with extreme care.

How Big Will It Be?

These computers do not need a special room with filtered air and raised floors. They work in an office-type environment, from ordinary 110V, 60-Hertz ac electricity. A bank of five cabinets, 72 inches high, 30 inches deep, and 100 inches wide overall will accommodate both of the computer systems mentioned in Table 1-1. The cabinets themselves will require 40 inches of access behind and in front. Wherever paper reports are to be printed, provision will have to be made for desktop or freestanding teletypewriters the size of large office typewriters. Usually, the output will be to video screens (the size of a portable television) that can be freestanding on desks, mounted in the panel, or custom-built into special consoles. The installation under discussion will have two teletypewriters and 10 video displays for the first (process) computer, and one teletypewriter and two video displays for the second (program development) computer.

How Powerful Is It?

The process system will be sufficiently powerful to scan all 320 inputs, calculate 1000 variables, output ten 200-field displays, and make adjustments to 24 valves (from the first scan to final adjustment) in less than five seconds— 8640 times a day, 365 days a (non-leap) year. The second system will be a powerful text-editing, program-compiling, system-development work station completely dedicated to such a purpose. Job turnaround time will be measured in minutes, and availability will be 100%.

Why Isn't the Trade Press Full of Success Stories?

Primarily because there have not been any outstanding success stories. There are three reasons: (1) misrepresentation of the machine's powers by the computer manufacturers, who have compounded their errors by misunderstanding the actual needs of the chemical process industries, (2) mismanagement and poor groundwork by the chemical and process industries in their embrace of this technology, an inability to grasp the task at hand, and (3) the confoundingly explosive growth of a confusing, complicated technology. The result has been far too few people actually knowing enough to do any project correctly, and far too many people thinking that they know enough to recommend, purchase, and install systems.

What Shouldn't I Expect?

Any increase in people-productivity for several years. The initial savings will be entirely in process material costs.

Where Can I Expect the Biggest Problems?

The biggest problems are in software development and in training people to get the most out of the system. Both of these problems will now be examined.

Problems in Software Development

There are four interrelated distractions to implementing computer-driven solutions. If they are not considered well enough beforehand, they can create an immobilizing confusion. Their interrelationship causes an overall impact much greater than their sum. The four distractions are: (1) rapidly declining hardware costs versus performance, (2) the low productivity in the developing and implementing of application software, (3) the order of magnitude in improvement which computers do offer over manual methods, and (4) the creation of software in "friendly" isolation.

Declining hardware costs brought about by the ever-increasing technical innovation of the microelectronic revolution has some of the attributes of an "arms race". Manufacturer A has on his drawing board a design that will be cheaper than the product manufacturer B is currently field testing. In turn, manufacturer B's product outperforms the device manufacturer C is just stepping up into quantity production, which in turn was technically superior to manufacturer D's original. Frighteningly, product evolution is measured in months, not years. Consequently, any decision made in buying hardware will be a wrong decision in just a few months. Some vendors talk of a "leading edge." This is nonsense. Real expertise only comes with the passage of time, and the passage of time makes the equipment obsolete. By definition, obsolete equipment cannot be "leading edge."

An early philosophy for any owner to adopt is whether the software effort be directed toward perfecting software for technically inferior equipment (compared to the current state-of-the-art), or toward keeping the software abreast of each and every technical improvement that comes along in hardware.

The low productivity of software development is partly people's inexperience, partly mechanical, partly ignorance of the workings within the machine, and partly the software itself.

Although simple programs work better and faster than complicated programs, the lesson is not learned. The initial attempts at writing a program always result in code that is too complex. It has to be broken down into smaller sections, made to work, and then chained together again. Although the actual choice of language will materially affect the development process, the choice is not always well made. The typing-in of the program is a tedious and slow process; debugging a program is involved and lengthy; printing out the listings and results is time-consuming. Overall, poorly constructed algorithms and ill-conceived program strategy can mean the difference between a compact, fast-running program operational in one day, and an unmitigated disaster that will not run even after a week of effort. The problem may be that the editing software is not powerful. It may be that the compiler lacks preprocessing and post-mortem analysis. Each individual person diddling with the software must learn by trial and error the way the processor works. Then each real-world application requires any lessons learned from previous applications to be used in different ways.

There are no obvious ways to overcome all these shortcomings. To recognize that these problems exist will direct effort to finding solutions for them.

The speed at which computers work distracts people because it is tempting to let the machine do this, do that, and do the other. After all, the thinking goes, it will use hardly any time at all. It is assumed that it is quicker for the machine to point out the errors than it is for an individual to look for them. This assumption is not always true. The machine is so fast that it can print *all* the results and let the individual make a visual selection. This approach is not as quick as it is thought to be. Reams of paper output are generated, then thrown away because small errors make the output valueless. Furthermore, the philosophy that only the relevant material will be kept and the rest discarded is not adhered to strictly.

Too few people appreciate the difference between software developed in "friendly" isolation and software field-proven in a "hostile" environment. Part of the problem lies in the fact that to mesh several programs in a system requires a knowledge of how each piece of hardware works. This knowledge is not required if only one program is run at a time. Sometimes a program that works perfectly in isolation cannot be run in conjunction with others. Friendly isolation also allows programmers to pursue super-elegant solutions for which implementation may not be practical. To get programs to work in hostile circumstances may require shoehorning various pieces of the program's features into small spaces—a very difficult challenge.

Last, poor hardware integration may not be apparent or it may be skirted easily in friendly systems. For systems in hostile environments, the problem must be faced head-on and solved. Good programs that fail in hostile environments are a source of many of the wiring and circuit changes that accompany the evolution of a computer product. Even the manufacturer is unable to foresee every last conflict or contention possible in his architecture.

The programming philosophy to follow is: the simpler, the shorter, the more dedicated a program; the faster the development, the quicker the execution, and the better the results.

Problems in Training

So much of so-called computer training (which, in the choice of words is symptomatic of the greater problem) is definitional and not user-orientated. The training is not geared to educating people to produce solutions efficiently by computer-assisted means. It teaches how to write code—not how to code efficient solutions.

There is a misconception that four weeks of training in "computers" will turn out computer-competent chemical engineers, each of whom will be able to use a system to the fullness of its bounds. Any such training program would be worthy of investigation if it even came close to its objectives. The overall complexity is such that, although competence can be achieved in a few months, complete mastery may not come in one year—if at all.

Recognition is not given to the fact that there are two different approaches to software. One is the uninformed black-box approach in which a user-ordered code is entered into the system by rote, and some time later an output becomes available. The other approach is when the innermost details of the machine's workings are understood and used. The code is optimized for the highest speed and minimum memory to produce efficacious user-oriented results at predictable times. Each approach has its place and time—one should not be confused with the other.

Many people are not prepared to put out the mental effort required for competence or mastery. The approach must be pedantic—it requires precisionists who emphasize minutiae. If people are not boggled by numbers, understanding the principles is easy. But to use those principles requires a mind that finds literal algebraic expressions no more confusing than arithmetic expressions, the ability to pursue abstract logic, and a quick memory for several hundred mnemonic devices, and details equal to a fair-sized book. All the foregoing must be in addition to a fully rounded chemical engineering skill that is technically proficient, people-competent, and business-oriented.

The creative process of writing a book is not confused with the mechanical process of typing a letter. Yet simple programming in a higher language is distinguished by very few people from the complex programming that can generate a system of software. The only people who should be accorded the title "programmer" are those capable of writing an assembly language routine for a higher-language program and adding a peripheral and its driver to a system. Many self-styled "programmers" are only finger painters. The emphasis in training must be on producing results and, if such results are not forthcoming, methods of modifying the software to get them. Software is a means, not an end. The training should be a recasting of chemical engineering knowledge as the computer allows the engineer to implement it.

The 16-Piece Process Computer Control Jigsaw (Figure 1-1)

1. Sufficient Economic Incentive

Are there sufficient benefits and savings to make a minicomputer installation financially viable?

2. Gaffer

Who will take the project on? Is there an engineer who already knows the process *and* a computer system's hardware and software? Will he accept the responsibility? Can he make the installation work? Needless to say, you are looking for a very rare person. He is indeed the keystone of the project.

3. The Make and Model of the Computer

Which one do you choose? (This choice is based on operating reliability, ease of maintenance, and range of compatible peripherals.) The CPU is the least costly of all the items in the project. A least-cost decision here may turn the project into an expensive mistake.

4. Computer Operating System

The operating system organizes the environment in which application programs run. It supervises the running of programs at requested times and handles the interfacing between programs/peripherals and programs/internal devices. An enhanced system will ensure that both the processor and the devices are used to their fullest capacity. Should a manufacturer's operating system be used? Should some other working system be duplicated? Should a new operating system be custom-written?

5. Computer Process Function

What will the computer do for the process? Will it collect data? Will it report condensed statistical analyses? Will it conduct process investigations? Will it raise alarms in an intelligent fashion? Will it be able to give energy

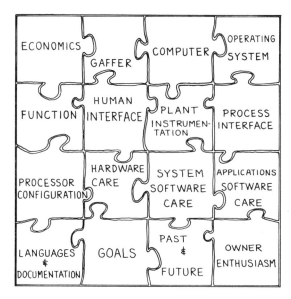

Figure 1-1. The process control computer jigsaw.

balances and material balances? Can it report management information? Will it conduct supervisory control of conventional instrumentation? Will it direct digital control with conventional instrumentation as a backup? Will there be only digital control and no conventional backup? Will the system be host to distributed microprocessor control?

6. Operator and Engineer Interface

How will process information be presented to the operators? Will it be through black-and-white CRTs, color CRTs, or only by printed reports? Will event-logging and typed reports be alternated on one printer or will there be a dedicated event-logging printer? Will any plotting be done? How are requests to the system to be handled? Will it be by dedicated one-strike function keys or by multistrike strings? Will prompting be by question-and-answer routines or by menus? Can the analog inputs and digital inputs/outputs be checked? Can the factors used in calculations be changed as instruments are reranged and as new instruments are installed? Can normal text with lower- and upper-case characters be entered into the machine?

7. Present Plant Instrumentation

Is it presently computer-compatible, easily made compatible, or incapable of being made compatible?

8.   Computer Process Interface

What speed, size, and complexity of process interface is required? Is it 30 points, 300 points, or 3000 points? Are the points to be measured once a second, once a minute, or once an hour? How many off/on switches are to be monitored? Is the machine expected to activate any switch closings/openings itself? How many control valves are to be supervised or operated? Are they moved by discrete digital pulses or by analog signals? Is the movement proportional to the number of pulses or to their duration? How fast can these pulses be transmitted? How many pulses does the equipment need to go from wide open to tightly shut?

9.   Processor Configuration

Is it to be stand-alone, host backup, or dual? Will the software development be done on the process computer in an off-line mode? Could it be done remotely on a much larger host computer? Is it proposed to install a second processor identical to the process processor for software development, debugging, and trials?

10.   Hardware

Who will do the development, the support, the trouble-shooting, and the maintenance?

11.   System Software

Who will do the development, the support, the trouble-shooting, and the maintenance?

12.   Applications Software

Who will do the development, the support, the trouble-shooting, and the maintenance?

13.   Languages and Documentation

What languages (plural) will be used? What languages (plural) will be supported? How much documentation is required? How will it be stored? Will there be paper or magnetic media vault copies?

14.   Goals

How will the goal-setting, performance measurement, and incentives be organized? What is the timetable? How will the improvement be measured?

15. Past and Future

Has anyone else ever controlled this process by computer? Has this computer been used for process control? Has the gaffer any particular expertise cogent to this project? What are the long-range future plans?

16. Owner Enthusiasm

Are the owner's heart and soul in controlling this process by computer?

Ginn's Rules

Rule 1. There is wisdom in the definition: *an expert is someone more than 30 miles from home.* Beware of outside experts who are flown in for meetings. Their knowledge is not necessarily proportional to distance traveled.* Should a computer "expert" say "sofware" instead of "software," raise your guard. His only expertise may be a mastery of buzzwords. Try to avoid meeting homegrown experts in their offices. Your being uncomfortable in the presence of stacks of listings, rows of impressive documentation binders, carelessly discarded printed circuit boards, and strange, whirring, blinking, beeping equipment gives the expert the psychological advantage. Remember that few computer experts are renaissance men—their knowledge, although deep, may be very narrow.

Rule 2. Programmers take simple arithmetic operations, precise logical statements, and straightforward algebra and produce unintelligible written procedures that neither man nor machine can follow immediately.

Rule 3. Documentation is an upper-management directive that serves only to cut programmers' productivity by 75%. Half of their productivity is lost when they write it, and a further half of their remaining productivity is lost when they read it.

Rule 4. Do not be panicked by the high cost of service calls. Remember how much you are paying your doctor and your automobile mechanic. The fee for emergency service is portal to portal. It includes travel time, smoke time, and meal breaks. The frustration of having a machine down, the apparent lack of activity while the fault is diagnosed, and the speed with which it is repaired just make it seem as though all you are paying for is traveling, phoning, smoking, and eating.

Rule 5. Self-documenting languages are unavailable because the compilers for the self-documenting languages are not documented well enough to release.

Rule 6. Software experts are a figment of the imagination.

*This rule produces unpredictable results south of southern latitude 60.

Rule 7. Like a first car and a first love, a person's first computer is a personality-molding experience.

Rule 8. How can you detect Boolean zero in a sales pitch? Generally, if the salesperson cannot demonstrate the product personally or answer your questions (but instead proffers assistance from his technical people), be sure that he will run out of technical help before he runs out of prospective customers or systems to sell.

Rule 9. Expect all vendor an supplier introductions to renewed frequently and expect contacts so made to be short-lived, particularly if your new-found contact is competent. The mushrooming demand for computers has put computer-literate people in short supply. The fact that computer-related skills are equally applicable in many divergent fields allows job-hopping right across sectors of the economy. Competent lower-echelon technicians will job-hop at six-month intervals. First, they will train themselves with the large electronic manufacturers, then look for an exciting application field, and finally work for that company which values their skills the most highly. Competent upper-echelon professionals also change jobs frequently, perhaps at two-year intervals. First, they round themselves out professionally, then they are tempted to move here and there by offers of ever-greater profit-sharing rewards. Executive burnout, motor vehicle accidents, and bankruptcy also contribute to rapid turnover.

Rule 10. No program works the first, second, or third time. If it does, there is something wrong with it.

Rule 11. Cussedness prevails. Despite all the expert skills and accumulated knowledge in the computer field, neither the organizations nor their products have been able to circumvent Murphy's law. In fact, the almost limitless combinations of almost identical items, each of which can go wrong, make the correct choice or a favorable result impossible on the first trial.

Rule 12. This rule has been removed temporarily. Field trials have shown poor reader acceptance. Minor problems with wording have to be resolved.

Rule 13. A computer invariably will fail catastrophically when being demonstrated.

Rule 14. Price is one scalar in a vector field. Delivery time is the other scalar, and it is more important than price.

Rule 15. In any discussions of what the computer can do, attack with the question, "What can *you* make the computer do?" This question puts everybody's feet back on the ground. The computer can do nothing without someone's making it do whatever it is that is to be done.

Rule 16. "To be released in . . . . " An overly optimistic statement made by a vendor who has something up his sleeve which he can't make work under all expected conditions. Add 12 months to his time, and purchase a product already on the market.

The Software Expense

Software people must strive to improve their productivity and become better at predicting what they can produce in one week. Current estimates are that it costs \$65 to develop one line of working code. Production is claimed to average only four lines of working code a day!

Is software an investment? Many people think it is, and base their actions on protecting it. It really is an expense and must be treated as such. No program should be attempted that cannot be working and paid out in six months. If people have ROI criteria on investments, they certainly should have to require payout horizons on software. That allows old programs (over six months) which are more than paid for to be thrown away without talk of a capital loss.

Source programs in higher languages must never ever be made machine-dependent; the whole advantage of a higher language (portability) will be lost. If this practice is adhered to, flexibility is maintained for the future. If current users of existing software on existing machines are happy, then new software written for new machines for new users is not precluded. Yet it somehow is always presented as an either/or choice. It is not. The old system and the new system can coexist and be run in parallel. Curiously, software will gravitate to the machine which provides the better compatibility. Remember: it is not the cost of the machine or the cost of existent software that is important. (Always get people to talk in terms of net costs for software, not gross costs.) The opportunity cost is the most important. The decisions must be directed toward giving programmers the capability of getting results in the quickest possible fashion.

If the performance of hardware is up and its price down, and if the performance of software is down and its price up—is an optimal balance possible? No! Too many developments are occurring. They succeed each other far too quickly for the earlier developments to be digested properly. But procrastination gives results that are worse. It is disadvantageous to seek out the "best buy"—it wastes too much time. The rewards are in application. The sooner they come, the better they are. The advice is: do not search for optimal results if that searching precludes procuring any results at all.

References

1. Skrokov, M.R., "The Benefits of Microprocessor Control," *Chemical Engineering*, October 11, 1976, pp. 133-139.

2. Seeman, R.C. and Nisenfeld, A.E., "Digital Distillation-Column Control Offers Flexibility," *Oil & Gas Journal*, October 20, 1975, pp. 57-60.

3. Kemp, D.W. and Ellis, D.G., "Computer Controls Fractionation Plants," *Oil & Gas Journal*, August 11, 1975, pp. 60-64.

2

PROCESS INSTRUMENTATION AND CONTROL

Microcontrol

The operation of a chemical process has two distinct parts: (1) the selection of optimum conditions for the process, and (2) the maintaining of the process at the chosen conditions. The former is the province of the operator and the production engineer using experience, calculations, and logged data to decide what conditions are optimum. The latter is the role of the instrument controllers. Controllers automatically maintain the measured variables at the chosen conditions. The digital computer/programmable processor can be used to replace either or both, although "to *supplement* either or both" is a better expression.

Let the overall process optimization by efficient allocation of resources and the choosing of conditions be macrocontrol. Let the maintaining of the equipment at the selected conditions be microcontrol. This chapter will describe fluid flow, some field measurements, and the traditional analog control theory that comprise microcontrol. This book will make brief mention of macrocontrol (see Chapter 8).

Fluid Flow

Microcontrol of a chemical process involves maintaining the temperatures, pressures, levels, and flows within predetermined limits of the chosen conditions. Obviously, in the case of flows (but less obviously in the other examples) such maintenance involves the regulation of a flow of mass, usually in liquid form (solids on a conveyor belt immediately spring to mind as an exception) to or from a vessel.

14

Whatever the so-called purpose of a controller is said to be, the primary objective is to control a fluid flow. Unfortunately, direct fluid flow control is given scant coverage in the literature. Most textbooks on process control begin with temperature control or level control. Yet to control the temperature of a column, the *flow* of steam to its reboiler must be varied. To control the level of liquid in a vessel, the *flow* of liquid either to or from the vessel must be regulated. To control the pressure of a header, a liquid *flow* either to or from the header must be manipulated. Consequently, there are several features of fluid flow that must be understood thoroughly before any control (analog or computer) is attempted. These features are: (1) what causes fluid to flow, (2) what reduces fluid flow, (3) how the fluid flow is regulated, and (4) how the fluid flow is measured.

A fluid flows from a high-pressure region toward a low-pressure region. If that fluid is contained in a pipe, then a pressure differential between the ends of the pipe will cause the fluid to flow from the high-pressure end to the low-pressure end. This difference in pressure can be caused by height (gravity flow), a straightforward difference in static pressures (as when liquid from a tank of high pressure is squirted into a tank of low pressure), or by mechanical means (such as a pump). The three can act in any combination, either supportive or appositive.

When a fluid flows in a pipe, it is subject to friction between itself and the walls of the pipe. Even in a pipe that has no artificial constriction, a given pressure differential can only push one certain maximum rate of flow through the pipe. At that equilibrium point, the frictional forces are equal and opposite to the force of the differential pressure across the ends of the pipe. Should some artificial adjustable constriction be placed in the line, then by increasing the constriction, the friction opposing the flow in the pipe could be increased and that maximum rate of flow reduced. Conversely, for any given restriction, the flow rate through the pipe can be increased by increasing the differential pressure.

As can be seen from the typical pump curves in Figure 2-1, as the head against which a pump must work increases, so the flow rate decreases. For any given pump casing, an increase in the size of the impeller (motor horsepower permitting) will result in either more flow for a given head, more head for a given flow, or some combination in-between. The actual flow available for a given head is a function of the particular casing, the impeller size, the motor size, and the motor speed.

For a given piping system of pipes, fittings, and valves in fixed positions, the head loss due to friction is proportional to the square of the flow. The actual head loss is a function of the fluid velocity, pipe diameter, length of the system, the number of fittings in the system, and the viscosity of the fluid (see Figure 2-2).

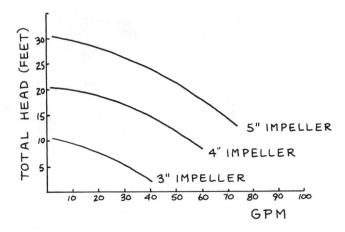

Figure 2-1. A typical pump curve—flow rate versus head pressure for various impeller sizes.

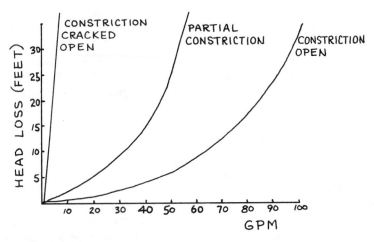

Figure 2-2. A typical piping system flow curve—flow rate versus head loss for various restrictions in the system.

For a given piping system and pump, a maximum flow rate exists for the combination (see Figure 2-3). As the adjustable constriction in the system is made more restricting, so the maximum rate of flow is reduced, the head differential available across the system is increased somewhat, and the head lost across the restriction is increased greatly. Finally, when the constriction is totally closed, the full head of the pump at no flow is across the restriction. This adjustable artificial constriction has served to regulate the flow. In fluid

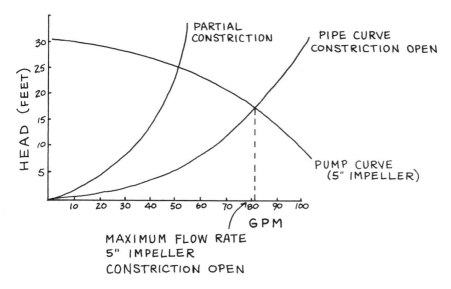

Figure 2-3. A pump curve with a piping system flow curve superimposed—flow rate versus head pressure and flow rate versus head loss.

flow it is called a control valve. There are three kinds of control valves: (1) disk control valves (ordinarily called "butterfly valves"), (2) ball control valves, and (3) plug control valves.

Disk control valves. These valves contain a flat disk. The diameter of the disk matches the inside diameter of the pipe containing the flow. When the disk is parallel to the line of flow, it offers hardly any restriction, and the fluid flow will be at its maximum. When the disk is at right angles to the direction of the flow of the fluid, it blocks the pipe completely, and there is no flow.

Ball control valves. This type of valve consists of spheres from which a large core has been taken. Depending on the particular construction, either the sphere's diameter can be the size of the line (and its cylindrical core much smaller), or the cylindrical core can be the inside diameter of the line (and the sphere's diameter much greater). In the former case, when the cylindrical core is in-line with the pipe, the flow of fluid in the pipe will be at its maximum, even though a restriction still exists in the line. In the latter case there will be virtually no restriction, and flow will be at the line's maximum. In either case, if the core of the sphere is turned at right angles to the pipe, the line of flow is blocked, and there will be no fluid flow.

Plug control valves.   There are many kinds of plug valve, but all are variations on positioning a cone or a cylinder in a circular hole (Figure 2-4). If the cone is completely withdrawn, the cross-sectional area of the hole will be at its maximum—there will be maximum flow. If the cone is advanced into the hole until the hole is completely blocked, there will be no flow (Figure 2-5).

Whatever kind of valve is in use, the required movement to close and to open can be provided by human muscle, an electrically driven screw mechanism, a hydraulic ram, or a pneumatically operated piston. Virtually all the control valves in chemical process control use a pneumatic piston drive (Figure 2-6). The overall distance that the operating mechanism travels from fully closed to fully open is called the stroke. With only one exception, there is not a linear relationship between the valve opening (as measured by percentage of stroke traveled) and actual fluid flow (as measured by percentage of maximum possible flow). (The one exception is a plug valve with a carefully configured cone-shaped plug that gives equal changes in flow for equal changes in stroke.)

Control Valve Curves

These curves are often presented in the literature in terms of the characteristics of the plug. The curve is the relationship between the percent of rated flow versus the percent of rated travel. The real world deviates from such idealized curves, sometimes quite considerably. The nature of the valve positioner, the strength of the valve spring, the process pressures against the plug itself, the friction losses in the piping system other than the control valve, the source and nature of the available head—all can and do affect the relationship between valve travel and rate of flow. Figure 2-7 portrays the real-world possibilities:

a.   An idealized linear relationship.

b.   An idealized "quick-opening" relationship where equal amounts of valve travel result in decreasing increments of flow.

c.   An idealized "equal-percentage" relationship where equal amounts of valve travel result in increasing increments of flow.

d.   A linear relationship where the valve has a weak-spring/early-opening positioner.

e.   A "quick-opening" relationship where the valve has a weak-spring/ early-opening positioner.

f.   An "equal-percentage" relationship where the valve has a weak-spring/ early-opening positioner.

Figure 2-4. Within control valves, varying sizes and shapes of openings in the stationery trim (which surround the moving plug) are available. Each shape is designed to give a control valve a characteristic which is best suited to the purpose of the valve. (Courtesy of Fisher Controls International Inc., Marshalltown, Iowa.)

Figure 2-5. An exploded view of a control valve showing the body, the trim, the plug, and the stem. (Courtesy of Fisher Controls International Inc., Marshalltown, Iowa.)

Figure 2-6. An assembled control valve with the pneumatic actuating cylinder assembly at the top connected by the stem to the plug within the body of the valve. (Courtesy of Fisher Controls International Inc., Marshalltown, Iowa.)

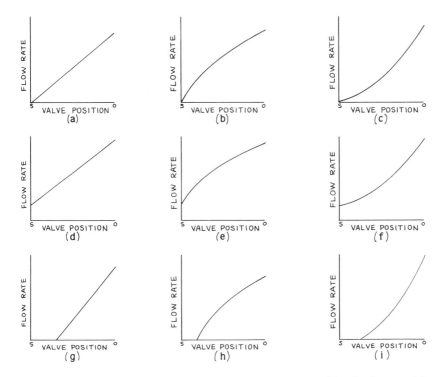

Figure 2-7. A series of valve curves—flow rate versus valve position for linear, quick-opening, and equal-percentage valves. For each valve, curves representing idealized, early-opening, and late-opening characteristics are presented.

g. A linear relationship where the valve has a strong-spring/late-opening positioner.

h. A "quick-opening" relationship where the valve has a strong-spring/late-opening positioner.

i. An "equal-percentage" relationship where the valve has a strong-spring/late-opening positioner.

If indeed the particular nature of the valve's travel/flow relationship can be discerned, it may not be repeatable because of changes in the fluid's properties. A rise of 100 degC in a liquid may reduce a piping system's friction losses by 40% (Figure 2-8a), allowing more flow for the same valve position. For gas flows, an overall rise in the piping system's pressure will cause more flow for the same valve position (Figure 2-8b).

There may be problems caused by improperly sized valves. The valve may be too small (Figure 2-9a). The valve may be too large (Figure 2-9b). The

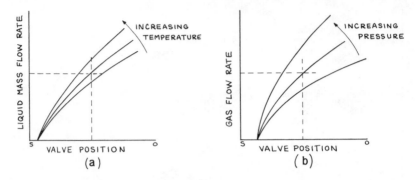

Figure 2-8. The variable nature of valve curves—flow rates vary at a fixed valve position because of varying conditions in the flowing liquid: (a) temperature (b) pressure.

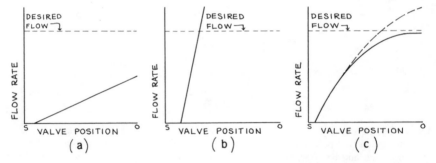

Figure 2-9. the valve curves of poorly sized valves: (a) too small, (b) too large, and (c) marginal (because the valve cannot pass the desired rates of flow under all possible conditions).

valve may be suitable in most circumstances but, under some circumstances of either reduced pressure (gases) or excessive viscosity (liquids), it may not be large enough (Figure 2-9c).

There are two singularly important non-linearities in control valve systems: the fixed rate of valve travel, and the acceleration forces on the fluid. Greater control actions will take proportionately longer than smaller control actions. This limitation is inherent in the system. Valve positioners do make the movement quicker, but even so the proportionality still exists. At low-flow rates, when the available head lost across the control valve is the greatest, the flow will respond the most quickly to control valve changes. At high-flow rates, when the available head lost across the control valve is the smallest, the flow will respond the most slowly to control valve changes. There may be an order of magnitude difference in response time between small changes at low-flow rates and great changes at high-flow rates—even if the changes are similar, percentage for percentage.

Fluid Flow Measurement

Currently, the bulk of fluid flow measurement in industry is done with orifice plates and differential pressure transmitters. Bernoulli's theorem states that for fluid flow through a horizontal Venturi (Figure 2-10),

$$\frac{v_1^2}{2g_c} + \frac{P_1}{\rho_1} = \frac{v_2^2}{2g_c} + \frac{P_2}{\rho_2} \tag{2-1}$$

where:

v = average velocity

P = fluid pressure

ρ = fluid density

g_c = gravitational constant

a = cross-sectional area

For an incompressible fluid, the volume flow is constant

$$v_1 a_1 = v_2 a_2 \tag{2-2}$$

and Equation (2-1) becomes

$$\frac{v_2^2}{2g_c} \left(1 - \left(\frac{a_2}{a_1} \right)^2 \right) = \frac{P_1 - P_2}{\rho} \tag{2-3}$$

With a knowledge of ρ, ΔP, and the two cross-sectional areas, the velocity can be calculated. From the velocity, the volume flow and the mass flow can be calculated

$$v_2 = \sqrt{\frac{2g_c}{\left(1 - \left(\frac{a_2}{a_1} \right)^2 \right)}} \cdot \sqrt{\frac{\Delta P}{\rho}}$$

$$V = v_2 a_2 = a_2 \sqrt{\frac{2g_c}{\left(1 - \left(\frac{a_2}{a_1} \right)^2 \right)}} \cdot \sqrt{\frac{\Delta P}{\rho}} \tag{2-4}$$

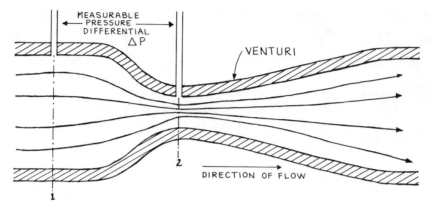

Figure 2-10. Flow in Venturi.

$$W = V \cdot \rho = a_2 \sqrt{\frac{2g_c}{\left(1 - \left(\frac{a_2}{a_1}\right)^2\right)}} \cdot \sqrt{\Delta P \cdot \rho} \tag{2-5}$$

where:

V = volume flow
W = mass flow

A Venturi is not a convenient apparatus for measuring flows in an industrial environment. The preferred apparatus is usually the orifice plate. As Figure 2-11 indicates, the cross-sectional areas of forward fluid flow at the points of pressure measurement are not well defined and certainly cannot be measured. Therefore, Equations (2-4) and (2-5) derived previously are used with a coefficient of discharge fudge factor and the orifice plate hole diameter. The coefficient incorporates the gravitational constant, the geometry constants, and a measurement unit conversion factor such that (even for inconsistent units)

$$V = d^2 \cdot C \cdot \sqrt{\frac{\Delta P}{\rho}} \tag{2-6}$$

$$W = d^2 \cdot C \sqrt{\Delta P \cdot \rho} \tag{2-7}$$

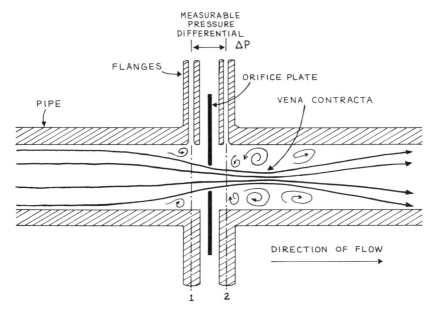

Figure 2-11. Flow through an orifice plate.

where:

d = orifice hole diameter

C = coefficient of discharge factor

Through many experiments, empirical value tables for C exist for any combination of flow rate, orifice plate hole, and pipe inside diameter. Because of the imperfect nature of the assumptions in the derivation, the basic equations can be refined greatly. Such refinements include allowances for temperature expansion of the orifice plate, Reynolds number corrections, liquid compressibility corrections, velocity of approach corrections, and flowing conditions.

The usual industrial practice for conventional instrumentation is to calculate a flow factor for the orifice plate (at the expected flowing conditions and range of the differential pressure transmitter) that includes all these refinements. It is then issued as a chart factor for the respective orifice plate/transmitter/flow recorder combination. For incompressible liquid flow, the flow equation does not attempt to correct for flowing density variations, although a flowing density value was assumed when the chart factor was calculated:

FLOW = OPF * DIVV

where:

FLOW = flow rate under consideration
   OPF = chart factor (orifice plate factor) in units of flow per chart division
  DIVV = the chart divisions of flow

For compressible gas flow, a variable gas expansion factor is included in the refinements, and a value for the flowing density at expected conditions is assumed when the chart factor is calculated. However, the flow calculation will include a flowing density correction factor:

$$\text{FLOW} = \text{OPF} * \sqrt{\text{GASFAC}} * \text{DIVV} \qquad (2\text{-}8)$$

where:

FLOW = flow rate under consideration
   OPF = chart factor (orifice plate factor) in units of flow per chart division at expected flowing conditions
  DIVV = the chart divisions of flow
GASFAC = $\dfrac{P_{AF}}{P_{EF}} * \dfrac{T_{EF}}{T_{AF}} * \dfrac{\%_{AF}}{\%_{EF}}$ = the flowing conditions corrections factor, the ratio of the expected flowing gas conditions (EF) to the actual flowing gas conditions (AF)

where:

P = absolute pressure
T = absolute temperature
$\%$ = purity

It is most important to remember that ΔP is measured in units of pressure (force per unit area), not feet of liquid. Force is the product of mass and acceleration. Consequently, in Bernoulli's theorem the "kinetic" term can only be related to the "potential" term if the density of the liquid is included. Obviously, mercury does not have to flow as rapidly as gasoline does to

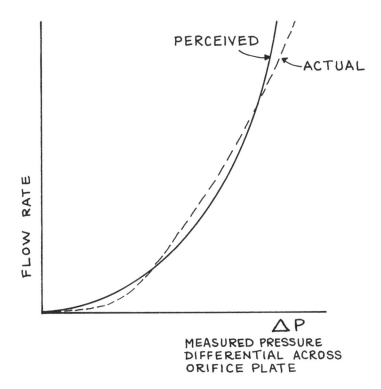

Figure 2-12. The variation between perceived and actual flows for various orifice plate pressure differentials.

produce an identical ΔP across an orifice plate. The density of the flowing liquid *must* be known whether the volume flow or the mass flow is being calculated. It is an erroneous and widespread belief that if only volume flow is to be measured the density need not be known. THAT IS WRONG.

That the coefficient of discharge is an empirically derived factor should not obscure the fact that it is itself a function of the flow rate. Unless the flow factor (or coefficient of discharge) is calculated anew for each flow rate, the flow will be overestimated at low and high rates and underestimated at medium rates (see Figure 2-12).

Non-Flow Measurement

It is also necessary to measure variables other than the differential pressure of flow to microcontrol the process. Temperature, pressure, level, pH, and concentration are examples.

Figure 2-13. A cutaway thermowell assembly showing the thermocouple inside. (Courtesy Leeds & Northrup Company.)

Temperature measurements are presented to the computer in one of two ways: either as a direct thermocouple emf, or as a transduced pressure or voltage signal. If it is a thermocouple signal (Figure 2-13), it will be necessary to know the cold-junction temperature to calculate correctly the hot-junction temperature. If a thermocouple is not practical or sufficiently accurate, then the temperature must be transduced. A liquid-filled bulb connected to a pressure transmitter or a resistance thermometer can be used for transduced measurements (Figure 2-14).

Pressure, differential pressure, and level ordinarily will be presented to the computer either directly as voltage ranges from electronic field transmitters (Figures 2-15 and 2-16), or indirectly as voltage from pneumatic/voltage transducers if pneumatic transmitters (Figure 2-17) are used in the field. All connections to the computer will be by twisted wire pairs terminating at

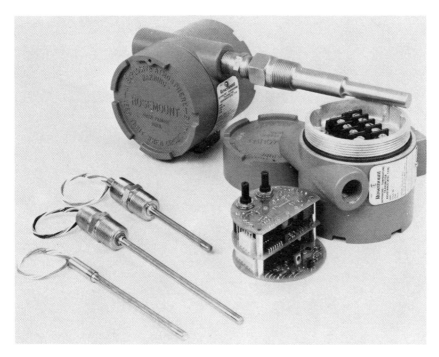

Figure 2-14. An electronic temperature transmitter, assembled and disassembled, showing various platinum resistance temperature sensors. (Courtesy of Rosemount, Inc.)

screw terminal blocks. Finally, if certain measurements cannot be presented to the computer as electric signals, their values can be entered manually by keyboard.

Controllers

In conventional instrumentation control valves are regulated by controllers (Figure 2-18). Controllers have either a manual or an automatic mode. A manual mode allows the operator to set and hold the control valve remotely at any desired opening. In automatic mode the controller will adjust the control valve automatically to keep some measured variable at some desired set point, the set point being selected by the operator. (See Figure 2-19 for a block diagram.) The programmable processor can be used either to set the set points on the controllers in automatic mode, or it can be used to set the positions of the control valves directly. The former is Set Point Supervisory Control (SSC). The latter is Direct Digital Control (DDC).

Figure 2-15. An electronic differential pressure transmitter with cover removed showing the span selector and topworks. (Courtesy The Foxboro Company.)

Figure 2-16. An assembled electronic differential pressure transmitter. (Courtesy of Rosemount, Inc.)

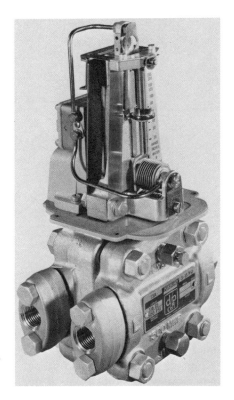

Figure 2-17. A pneumatic differential
pressure transmitter with cover removed.
(Courtesy The Foxboro Company.)

Figure 2-18. A panelboard-mounted conventional analog
controller. (Courtesy The Foxboro Company.)

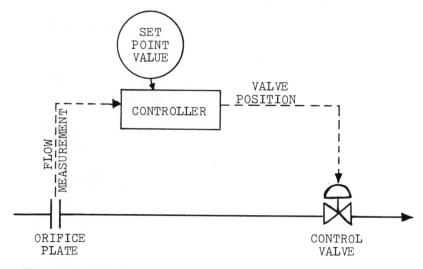

Figure 2-19. A block diagram for an analog controller for fluid flow control.

For every controller connected to the computer, at least five connections must be made. The first is the remote/local switch. At "switch open," the computer is to monitor and cannot intervene. At "switch closed," the computer can make adjustments if necessary. The second connection is the position of the controller's set point. The third connection permits the computer to increase the set point; the fourth connection permits the computer to decrease the set point. The last connection informs the controller whether or not the computer is in an operational condition (Figure 2-20).

Conventional PID Algorithm

For analog controllers, the proportional-integral(reset)-derivative algorithm has been the traditional approach. It is written as a differential equation

$$dp = K_p * e + K_R * \int edt + K_D * \frac{de}{dt}$$

where:

e = instantaneous error between set point (desired value) and measurement (actual)

$\int edt$ = the sum of the errors past with respect to time

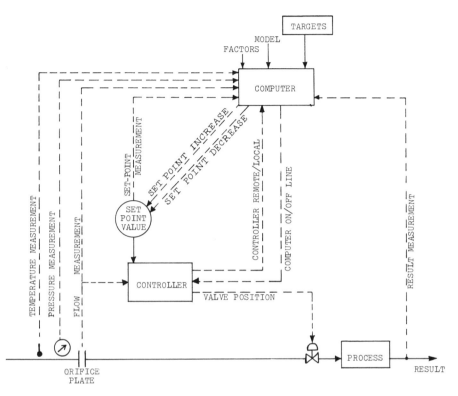

Figure 2-20. A block diagram for a computer-supervised analog controller for fluid flow control.

$\dfrac{de}{dt}$ = the rate at which the error is increasing/decreasing with respect to time

K_p = the PROPORTIONAL factor (applies only to proportional action)

K_R = the RESET factor (applies only to reset action)

K_D = the DERIVATIVE factor (applies only to derivative action)

dp = the incremental control action

The terms have been separated deliberately to conform with Chapter 8.

The PROPORTIONAL Term

The proportional factor is defined in terms of percent. Proportional action is corrective action of the valve position that is directly proportional to the deviation between the actual value and the desired value of the controlled variable. If the deviation of the controlled variable is expressed as a fraction of the desired value of the controlled variable, and if the corrective action is expressed as a fraction of the desired value of the valve position, then the correspondence between the two fractions expressed as a percent is the proportional band

$$\text{Proportional band \%} = \frac{\text{Error}}{\text{Desired value}} * \frac{\text{Valve position}}{\text{Correction}} * 100$$

At a 50% band, the valve-movement fraction is twice the value-movement fraction. At a 400% band, the valve-movement fraction is one-quarter of the value-movement fraction. Therefore, a move of 1% of the flow from its desired position causes the valve to move one-quarter percent from its position.

The RESET Term

The reset term is defined as a time period. The reset action is that control action that removes the offset between the actual value and the desired value of the controlled variable. Proportional action will occur only if a deviation in the controlled variable exists. Its action is to return the valve to its original position. For proportional action to maintain the valve at its new position, there must be a permanent offset between the actual and desired values of the controlled variable. Reset action moves the valve slowly but consistently until that offset is removed. It is defined as the time required for the controller to remove an offset of some given original size.

The DERIVATIVE Term

Derivative action is that control action that is proportional to the rate at which the error term is increasing or decreasing. It is rarely used in conventional analog controllers.

Analog controllers must be field-tuned to the best performance for a given set point. The proportional and reset settings chosen will be symmetrical and linear in action for measurement-value deviations above and below the set point. The settings will not vary automatically if the set point itself is moved up or downscale. Analog controllers cannot accommodate dead-time at all. Their control action is based on the current error measurement, which is not the result of the current control action.

3

CASE STUDY—
THE WITCHES' BREW PROCESS

Process Summary

In the Witches' Brew Process the raw materials Double, Bubble, Toil and Trouble are heated to a high temperature inside a reactor while being agitated vigorously. This mixture is held at a high temperature and pressure in a soaker for a period of time and then flashed into a flash drum where it cools rapidly. The gas, Noxious, is vented from the flash drum to a flare system, where it is burned into harmless combustion products.

The cooled liquid mixture flows from the flash drum through a mixer to the neutralizer. The final ingredient, Mare's Sweat, is added in the mixer to neutralize the liquid. It is a slow reaction, but does go to completion in the neutralizer. The final product, Witches' Brew, has been formed, but it exists in an aqueous state with various impurities and a by-product that serves as an effective paint remover, EPR.

The Witches' Brew, EPR, and some of the impurities are removed from the water by solvent extraction with Frogs' Spit. The unwanted water layer is discarded to waste disposal. The Witches' Brew and EPR are recovered from the Frogs' Spit by solvent evaporation in the evaporator. The evaporated solvent is recovered and returned to the solvent extraction process. The Witches' Brew and EPR mixture is stored temporarily in the crude product tank before being separated by distillation in the distillation column. EPR is the overhead product. Witches' Brew is the bottoms product. Tincture of TSJ is added to the distillation column feed. It suppresses the thermal decomposition of Witches' Brew during distillation. Each product stream accumulates in its own production run tank. On a daily basis, each production run tank's contents are transferred to its respective shipping tank.

35

See Table 3-1 for the production balance, and Table 3-2 for the process economics. Figure 3-1 is a process flow sheet of the Witches' Brew Process. Figure 3-2 gives a detailed material and energy balance for the Witches' Brew Process.

Equipment Inventory

Tanks

T-01   Bubble raw material storage tank
T-02   Toil and Trouble raw material storage tank
T-03   Mares' Sweat raw material storage tank
T-04   TSJ Tincture additive storage tank
T-05   Frogs' Spit solvent storage tank
T-06   Crude product storage tank
T-07   EPR product run tank
T-08   Witches' Brew product run tank
T-09   EPR shipping tank
T-10   Witches' Brew shipping tank

Special vessels

V-01   Reactor
V-02   Soaker
V-03   Neutralizer
V-04   Evaporator
V-05   Distillation column

Drums

D-01   Flash drum
D-02   First extractors drum
D-03   Second extractors drum
D-04   Third extractors drum
D-05   Distillation reflux drum

Heat exchangers

E-01   Reactor heater
E-02   Flash drum vapor condenser
E-03   Flash drum liquid cooler
E-04   Evaporator condenser
E-05   Evaporator reboiler
E-06   Distillation column condenser
E-07   Distillation column reboiler

Table 3-1
Process-Material Balance

Production:			
Witches' Brew			24 million pounds/year
EPR			10 million pounds/year
Raw Material Consumption:			
Double		80% yield	5 million pounds/year
Bubble	50% aqueous solution	85% yield	30 million pounds/year
Toil & Trouble	70% aqueous solution	90% yield	12 million pounds/year
(5/7 mix)			
Mares Sweat			3 million pounds/year
TSJ Tincture			24 thousand pounds/year
Utilities:			
Steam			102 million pounds/year
Cooling water			470 million gallons/year
Electricity			2 million kWh/year
Process Manpower:			
Shiftworkers	8		
Day workers	3		
Maintenance	2		
Plant manager	1		
Total:	14 persons		

Mixers

M-01   Mares' Sweat static mixer
M-02   Neutralizer agitator

Pumps and motors

P-01   Bubble feed pump
P-02   Toil and Trouble feed pump
P-03   Reactor recycle pump
P-04   Mares' Sweat feed pump
P-05   Extractors feed pump
P-06   First extractors pump
P-07   Second extractors pump
P-08   Third extractors pump
P-09   Extractors raffinate pump
P-10   Extractors solvent pump
P-11   Evaporator recycle pump
P-12   Distillation column feed pump
P-13   TSJ Tincture pump
P-14   Distillation column recycle pump
P-15   Distillation column reflux pump
P-16   EPR run tank pump

Table 3-2
Process Economics (all figures $M)

Capital

Direct fixed capital		3000
Pro-rata share utility capital		474
Pro-rata share other, administrative, and sales capital		100
Working capital (half month's raw material supply)		114
Working capital (one month's product supply)		362
	Total Capital	4050

Annual Costs
  Variable

Raw materials:	Double, 5 million pounds @ 5¢/#		250
	Bubble, 30 million pounds @ 6¢/#		1800
	Toil & Trouble, 12 million pounds @ 2¢/#		240
	Mares' Sweat, 30 million pounds @ 15¢/#		450
		Total Raw Material Costs	2740
Utilities:	steam, 102 million pounds @ $3/M#		306
	water, @ $1/10,000 g		47
	electricity, @ $12/MKw		24
		Total Utilities	377
Labor:	14 people at $22,000/yr		308
Miscellaneous supplies (including TSJ Tincture)			298
		Total Variable Costs	3723

  Fixed

Maintenance @ 4% of DFC		120
Depreciation @ 10% of DFC		300
Taxes and Insurance @ 2% of TC		81
Administrative & Factory Expense @ 3% of TC:		121
	Total Fixed Costs	622
	Total Annual Costs	4345

Sales:

Witches' Brew, 24 million pounds @ 18¢/#		4320
EPR, 10 million pounds @ 10¢/#		1000
	Total Sales	5320

Profitability:

Total Sales	5320
Total Costs	4345
Selling Expense	102
Operating Margin	873

$$\text{ROIBT} = \frac{\text{OP MARG}}{\text{TOT CAP}} = \frac{873}{4050} = 21.6\%$$

Return on investment 21.6% before taxes

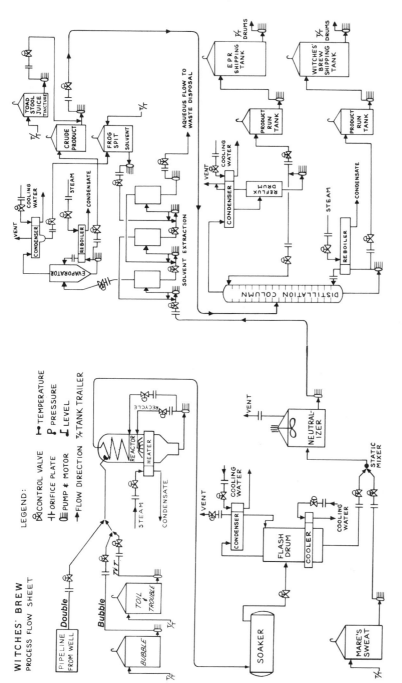

Figure 3-1. The process flow sheet of the Witches' Brew Process.

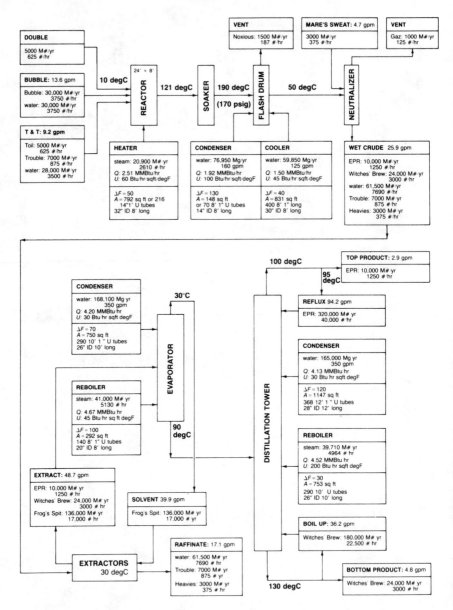

Figure 3-2. The material and energy balance of the Witches' Brew Process.

P-17 EPR shipping tank pump
P-18 Witches' Brew run tank pump
P-19 Witches' Brew shipping tank pump

Tank level indicators (locally mounted in field)

LI-001 level, Bubble storage tank (T-01)
LI-002 level, Toil and Trouble storage tank (T-02)
LI-003 level, Mares' Sweat storage tank (T-03)
LI-004 level, TSJ Tincture storage tank (T-04)
LI-005 level, Frogs' Spit storage tank (T-05)
LI-006 level, crude storage tank (T-06)
LI-007 level, EPR run tank (T-07)
LI-008 level, Witches' Brew run tank (T-08)
LI-009 level, EPR shipping tank (T-09)
LI-010 level, Witches' Brew shipping tank (T-10)

Tank temperature indicators, (locally mounted in field)

TI-011 temperature, Bubble storage tank (T-01)
TI-012 temperature, Toil and Trouble storage tank (T-02)
TI-013 temperature, Mares' Sweat storage tank (T-03)
TI-014 temperature, TSJ Tincture storage tank (T-04)
TI-015 temperature, Frogs' Spit storage tank (T-05)
TI-016 temperature, crude storage tank (T-06)
TI-017 temperature, EPR run tank (T-07)
TI-018 temperature, Witches' Brew shipping tank (T-08)
TI-019 temperature, EPR run tank (T-09)
TI-020 temperature, Witches' Brew shipping tank (T-10)

Pressure indicators (locally mounted in field)

PI-021 discharge pressure, Bubble feed pump (P-01)
PI-022 discharge pressure, Toil and Trouble feed pump (P-02)
PI-023 discharge pressure, reactor recycle pump (P-03)
PI-024 discharge pressure, Mares' Sweat feed pump (P-04)
PI-025 discharge pressure, extractors feed pump (P-05)
PI-026 discharge pressure, first extractors pump (P-06)
PI-027 discharge pressure, second extractors pump (P-07)
PI-028 discharge pressure, third extractors pump (P-08)
PI-029 discharge pressure, extractors raffinate pump (P-09)
PI-030 discharge pressure, extractors solvent pump (P-10)
PI-031 discharge pressure, evaporator recycle pump (P-11)
PI-032 discharge pressure, distillation column feed pump (P-12)
PI-033 discharge pressure, TSJ Tincture pump (P-13)
PI-034 discharge pressure, distillation column recycle pump (P-14)

PI-035   discharge pressure, distillation column reflux pump (P-15)
PI-036   discharge pressure, EPR run tank pump (P-16)
PI-037   discharge pressure, EPR shipping tank pump (P-17)
PI-038   discharge pressure, Witches' Brew run tank pump (P-18)
PI-039   discharge pressure, Witches' Brew shipping tank pump (P-19)
PI-040   discharge pressure, static mixer, flash drum feed
PI-041   discharge pressure, static mixer, Mares' Sweat feed
PI-042   discharge pressure, static mixer, downstream

Process level transmitters (field mounted)
Recorders (panelboard mounted)

LT-043, LR-043   level, reactor (V-01)
LT-044, LR-044   level, flash drum (D-01)
LT-045, LR-045   level, neutralizer (V-03)
LT-046, LR-046   level, first extractors drum (D-02)
LT-047, LR-047   level, second extractors drum (D-03)
LT-048, LR-048   level, third extractors drum (D-04)
LT-049, LR-049   level, evaporator (V-04)
LT-050, LR-050   level, evaporator condenser (E-04)
LT-051, LR-051   level, distillation column (V-05)
LT-052, LR-052   level, reflux drum (D-05)

Process pressure transmitters (field mounted)
Recorders (panelboard mounted)

PT-053, PR-053   pressure, Double feed flow
PT-054, PR-054   pressure, reactor top (V-01)
PT-055, PR-055   pressure, soaker (V-02)
PT-056, PR-056   pressure, flash drum condenser (E-02)
PT-057, PR-057   pressure, first extractors drum (D-02)
PT-058, PR-058   pressure, distillation column bottom (V-05)
PT-059, PR-059   pressure, distillation column top (V-05)

Process temperature sensors and transmitters (field mounted)
Recorders (panelboard mounted)

TS-060, TT-060, TR-060   temperature, Double feed flow
TS-061, TT-061, TR-061   temperature, reactor bottom (V-02)
TS-062, TT-062, TR-062   temperature, flash drum condenser liquid out (E-02)
TS-063, TT-063, TR-063   temperature, flash drum forward flow (D-01)
TS-064, TT-064, TR-064   temperature, evaporator bottom (V-04)
TS-065, TT-065, TR-065   temperature, evaporator condenser liquid out (E-04)

TS-066, TT-066, TR-066   temperature, distillation column condenser liquid out (E-06)

TS-067, TT-067, TR-067   temperature, distillation column bottom (V-05)

Orifice plates, flow transmitters (field mounted)
Recorders (panelboard mounted)

OP-068, FT-068, FR-068   flow, Double feed
OP-069, FT-069, FR-069   flow, Bubble feed
OP-070, FT-070, FR-070   flow, Toil and Trouble feed
OP-071, FT-071, FR-071   flow, reactor cycle
OP-072, FT-072, FR-072   flow, reactor forward
OP-073, FT-073, FR-073   flow, reactor heater steam
OP-074, FT-074, FR-074   flow, flash drum condenser vent
OP-075, FT-075, FR-075   flow, flash drum condenser water
OP-076, FT-076, FR-076   flow, flash drum cooler water
OP-077, FT-077, FR-077   flow, flash drum forward
OP-078, FT-078, FR-078   flow, Mares' Sweat feed
OP-079, FT-079, FR-079   flow, neutralizer vent
OP-080, FT-080, FR-080   flow, extractors feed
OP-081, FT-081, FR-081   flow, extractors solvent
OP-082, FT-082, FR-082   flow, extractors raffinate
OP-083, FT-083, FR-083   flow, extractors extract
OP-084, FT-084, FR-084   flow, evaporator recycle
OP-085, FT-085, FR-085   flow, evaporator bottoms
OP-086, FT-086, FR-086   flow, evaporator overhead
OP-087, FT-087, FR-087   flow, evaporator condenser water
OP-088, FT-088, FR-088   flow, evaporator reboiler steam
OP-089, FT-089, FR-089   flow, TSJ Tincture feed
OP-090, FT-090, FR-090   flow, distillation column feed
OP-091, FT-091, FR-091   flow, distillation column bottoms
OP-092, FT-092, FR-092   flow, distillation column forward
OP-093, FT-093, FR-093   flow, distillation column reflux
OP-094, FT-094, FR-094   flow, distillation column condenser water
OP-095, FT-095, FR-095   flow, distillation column reboiler steam
OP-096, FT-096, FR-096   flow, distillation column recycle

Controllers (panelboard mounted)
Control valves (field mounted)

Example: index number: title
                                controlled variable
                                control valve

C-097, CV-097:   Double feed flow control
                 Double feed O/P
                 Double feed C/V
C-098, CV-098:   Bubble feed flow control
                 Bubble feed O/P
                 Bubble feed C/V
C-099, CV-099:   Toil and Trouble feed flow control
                 Toil and Trouble feed O/P
                 Toil and Trouble feed C/V
C-100, CV-100:   reactor recycle flow control
                 reactor recycle O/P
                 reactor recycle C/V
C-101, CV-101:   reactor level control
                 reactor level
                 reactor forward flow C/V
C-102, CV-102:   reactor recycle temperature control
                 reactor recycle temperature
                 reactor heater steam flow C/V
C-103, CV-103:   soaker pressure control
                 soaker pressure
                 soaker forward flow C/V
C-104, CV-104:   flash drum pressure control
                 flash drum condenser pressure
                 flash drum vent C/V
C-105, CV-105:   flash drum condenser temperature control
                 flash drum condenser liquid temperature
                 flash drum condenser water C/V
C-106, CV-106:   flash drum temperature control
                 flash drum forward flow temperature
                 flash drum cooler water C/V
C-107, CV-107:   flash drum level control
                 flash drum level
                 flash drum forward flow C/V
C-108, CV-108:   neutralizer pH control
                 neutralizer pH
                 Mares' Sweat C/V
C-109, CV-109:   neutralizer level control
                 neutralizer level
                 extractors feed C/V
C-110, CV-110:   first extractors drum level control
                 first extractors drum level
                 first extractors drum C/V

C-111, CV-111: second extractors drum level control
second extractors drum level
second extractors drum C/V

C-112, CV-112: third extractors drum level control
third extractors drum level
third extractors drum C/V

C-113, CV-113: extractors pressure control
first extractors drum pressure
extract C/V

C-114, CV-114: solvent feed flow control
solvent feed O/P
solvent C/V

C-115, CV-115: evaporator level control
evaporator level
evaporator bottoms C/V

C-116, CV-116: evaporator temperature control
evaporator bottoms temperature
evaporator reboiler steam C/V

C-117, CV-117: evaporator condenser temperature control
evaporator condenser liquid temperature
evaporator condenser water C/V

C-118, CV-118: evaporator condenser liquid level control
evaporator condenser liquid level
evaporator forward flow C/V

C-119, CV-119: TSJ Tincture feed flow control
TSJ Tincture feed O/P
TSJ Tincture feed C/V

C-120, CV-120: distillation column feed flow control
distillation column feed flow O/P
distillation column feed C/V

C-121, CV-121: distillation column level control
distillation column bottoms level
distillation column bottoms flow C/V

C-122, CV-122: distillation column reflux drum level control
distillation column reflux drum level
distillation column forward flow C/V

C-123, CV-123: distillation column reflux flow control
distillation column reflux O/P
distillation column reflux C/V

C-124, CV-124: distillation column condenser temperature
control
distillation column condenser liquid

	temperature
	distillation column condenser water C/V
C-125, CV-125:	distillation column bottoms temperature control
	distillation column bottoms temperature
	distillation column reboiler steam C/V

Integrators (panelboard mounted)
I-126 integrator, Double feed flow
pH sensor and transmitter (field-mounted)
Recorder (panelboard-mounted)
pH T-127, pH R-127 neutralizer pH
Temperatures, Fe/Constantin thermocouples (field-mounted)
Multipoint recorders (panelboard-mounted)

TR-128	TR-129
#1 bottom reactor	#1 distillation column overhead
#2 middle reactor	#2 distillation column condenser vent
#3 top reactor, vapor space	#3 distillation column condenser liquid
#4 reactor forward flow to soaker	#4 tray 1
#5 first quarter soaker	#5 tray 2
#6 second quarter soaker	#6 tray 3
#7 third quarter soaker	#7 tray 4
#8 final quarter soaker	#8 tray 5
#9 flash drum bottom	#9 tray 6
#10 flash drum top	#10 tray 7
#11 flash drum condenser liquid	#11 tray 8
#12 flash drum condenser vent	#12 tray 9
#13 neutralizer	#13 tray 10
#14 extractor feed	#14 tray 11
#15 extractor solvent	#15 tray 12
#16 extractor raffinate	#16 tray 13
#17 extractor extract	#17 tray 14
#18 evaporator overhead	#18 tray 15
#19 evaporator condenser vent	#19 distillation column reboiler out
#20 evaporator reboiler out	#20 spare

Process and Instrumentation Description

The raw materials for Witches' Brew are Double, Bubble, and Toil and Trouble. Aqueous solutions of Bubble and Toil and Trouble are brought to the plant in tank trailers and stored in their respective storage tanks. The nominal strength of the Bubble solution is 50%; it varies ± 2%, shipment to shipment. The nominal strength of the Toil and Trouble solution is 30%, but it is subject to greater fluctuations in purity than is Bubble. Past experience has shown that the actual strength ranges between 25% and 37%. The proportion of the mix is also a nominal figure. Although specified to be 5:7, it does range from 6:6 to 4:8. The gas, Double, is piped directly into the plant from its production well and varies in purity, temperature, and pressure.

Double, Bubble, and Toil and Trouble are flow-controlled to the reactor. Double and Toil and Trouble are fed in strict ratio proportion to Bubble. Within the reactor, the raw materials are heated and violently agitated. They react to form the product Witches' Brew and the by-product EPR. Some thermal decomposition results in the production of a gas, Noxious. There are some side reactions that produce an acid and other undesirable impurities. The reactor's contents are sprayed over the heating tubes in a continuous recycle. This gives heat transfer and excellent mixing. The recycle rate itself can be varied by adjustment of a flow-control valve in the recycle line.

On their introduction to the top of the reactor, the raw materials are sprayed over a bundle of tubes which contain the reactor product forward flow. The cross-exchange of heat initiates the vapor phase of the reaction. The liquid drops and spray fall to the bottom of the reactor, where the higher temperatures initiate the liquid phase portion of the reactions. The reactor bottom temperature is maintained at 121 degC by a temperature controller acting on the steam flow-control valve in the steam line to the reactor heater. The reactor level is maintained by a level controller acting on the flow-control valve in the reactor forward-flow line to the soaker. The liquid phase reactions go to completion in the soaker. The exothermic reaction raises the soaker exit temperature to 190 degC. The soaker is maintained at 170 psig by a back-pressure controller acting on a flow-control valve in the soaker exit line. The high-temperature, high-pressure flow from the soaker is flashed through that pressure-control valve into the flash drum. This flash drum is maintained at 30 psig by a back-pressure controller acting on a flow-control valve in the vent line from the flash drum condenser.

In the cooler and the condenser of the flash drum the reaction products are cooled down to 50 degC. The condenser has a temperature controller acting on a flow-control valve in the cold-water line to the condenser. The cooler has a temperature controller acting on a flow-control valve in the cold-water line to the cooler. The condensable vapors are returned to the flash drum. The non-condensable gas, Noxious, is vented (as it accumulates) through the pressure-control valve to a flare system where it is burned. The liquid contents of the flash drum are cooled by submerged water-cooled coils. The cooled flash drum liquid is pressured through a static mixer to a neutralizer. The level in the flash drum is level-controlled by the flash drum forward flow. In the mixer the cooled, wet, impure Witches' Brew is completely mixed with Mares' Sweat, which is flow-controlled in strict proportion as it is pumped from the Mares' Sweat storage tank. The Mares' Sweat neutralizes the acidic crude product to a pH between 6.8 and 8.1. With some agitation in the neutralizer, the acid neutralization goes to completion to form Gaz, water, and Heavies. The neutralizer also acts as a bulk storage tank and holds the neutral aqueous solution of Witches' Brew, EPR, Trouble, Heavies, and some impurities prior to its being fed to the three-stage puri-fication process.

The purification process is solvent extraction of Witches' Brew and EPR from the aqueous solution by Frogs' Spit; evaporation of the Frogs' Spit from the Witches' Brew/EPR mixture; and finally separation by distillation, the EPR being taken overhead and the Witches' Brew being left as the column's bottom product. The extractors are a three-stage, counter-current process. A 2:3 volume ratio of wet crude to solvent has been found to be effective. On a dry basis, this is a 4:1 ratio of solvent to solute. Greater solvent ratios lead to unacceptable solvent losses in the raffinate and greater utility consumption in the evaporator. Lesser solvent ratios lead to a loss of product in the raf-finate. The 2:3 ratio is an acceptable compromise between the two extremes. The exit temperatures of the flash drum and evaporator condenser are closely monitored to prevent the extractor's temperature from rising above 55 degC.

In the solvent extraction process the neutral aqueous mix is flow-controlled to the first mixing pump. It is mixed furiously in the pumped recycle loop with the solvent layer from the top of the second drum. The mixture is settled in the first drum. The lower, heavier aqueous layer from the first drum is level-controlled to the second mixing pump. Again, it is mixed violently in a pumped recycle loop, but this time with the solvent layer from the top of the third drum. The mixture is settled in the second drum. The lower, heavier aqueous layer from the second drum is level-controlled to the third mixing pump. This time the lower layer is mixed thoroughly in the pumped recycle loop with fresh Frogs' Spit solvent. The fresh solvent and second drum aqueous layer mix are settled in the third drum. The lower, heavier aqueous layer from the third drum is pumped through a level-controlled control valve to waste disposal. No attempt is made to recover Trouble. All the Witches' Brew and EPR have been extracted from the aqueous layer by the Frogs' Spit solvent.

The Frogs' Spit is pumped from the solvent storage tank through a flow-controlled control valve to the third mixing pump. It settles out as the lighter, upper layer in the third drum. It is pressured to the second mixing pump and thence to the second drum. It settles out as the lighter, upper layer in the second drum. It is pressured from there to the first mixing pump where it mixes with the feed and thence to the first drum. It settles out as the lighter, upper layer. It is pressured out of the first drum through a pressure-controlled control valve to the evaporator.

It should be noted that the aqueous stream gets successively weaker in Witches' Brew and EPR as it flows from the first mixer to the first drum to the second mixer to the second drum to the third mixer to the third drum and out; whereas the solvent gets progressively richer in Witches' Brew and EPR as it flows from the third mixer to the third drum to the second mixer to the second drum to the first mixer to the first drum and out. In the evaporator the Frogs' Spit and Witches' Brew/EPR mixture are separated by evaporating the Frogs' Spit. The evaporator bottoms are heated to 90 degC and the evaporator is operated at atmospheric pressure. Under these conditions, virtually all the Frogs' Spit is evaporated. The Frogs' Spit is evaporated, condensed overhead, and then gravity-flowed through a level-controlled flow-control valve to the solvent feed tank. The evaporator bottoms are maintained at the desired temperature by a temperature controller acting on a flow-control valve in the steam line to the evaporator reboiler. The hot evaporator bottoms are pumped to the crude product tank. The rate of flow is controlled by the evaporator bottoms level controller acting on a flow-control valve in the line to the crude product tank. The crude product tank serves as a storage buffer for the feed to the 15-tray distillation column.

In the distillation column the EPR is taken overhead and Witches' Brew is the column's bottom product. The greatest possible purity is sought in both salable products; consequently, the tower is run at a very high reflux ratio. The temperature of the bottom tray controls the stream flow to the reboiler. The bottoms temperature is maintained at 130 degC. The column's bottom flow is level-controlled out to the Witches' Brew run tank. The bottoms recycle flow through the reboiler is recorded. To inhibit thermal degradation of the products in the column, tincture of Toadstool Juice is metered into the column feed. If there is insufficient tincture, the aldehyde content of the EPR increases. With too much tincture, the color of the Witches' Brew is adversely affected. The reflux to the column is flow-controlled from the reflux drum to the top tray. The overhead forward flow is level-controlled from the reflux drum to the EPR product run tank. The cooling water rate to the condenser is set to maintain a liquid out temperature of 95 degC. The condenser runs at atmospheric pressure and consequently the top tray temperature is approximately 100 degC. The top product and the bottom product are collected over 24-hour periods in their respective run tanks.

Each product's quality is checked before transfer to its shipping tank. If the particular run tank is of too low a quality it can either be returned to the crude product tank or to the neutralizer. The actual choice is determined by the nature of the quality problem. Both products are sold in bulk in tank trailer quantities and packaged in 55-gallon drums.

4

BASIC COMPUTER HARDWARE

Binary Representation

Binary states are characterized by duality. Consider the following: a bulb, a switch, a toroidal magnetic cell, and a capacitor. Each of these can have only one of two states at any one time. A bulb can be either glowing (on) or unlit (off). A switch can cause a circuit to be closed (on) or open (off). A toroidal magnetic cell can be clockwise (on) or counterclockwise (off). A capacitor is charged (on) or discharged (off). If we define on as a 1 and off as a 0, then a lit bulb is 1, an open switch is 0, a clockwise toroidal magnetic cell is 1, and a discharged capacitor is 0.

If we were to generalize this concept further, we could associate such numeric descriptors as 0 and 1 to certain electric circuit attributes: in particular, current and voltage. Because both exist in a continuum, the definition must be restrictive. As an example, for "current," the definitions will be of the form: a current between zero and 100 microamps will be 0, and a current between 50 and 200 milliamps will be 1. Currents outside these limits are undefined. In like fashion, for voltage: a voltage between 0.0 and 0.3 volts will be a 0, and a voltage between 2.8 and 3.1 volts will be a 1. Voltages outside these limits are undefined.

Binary information is knowledge that is coded in binary form. It can be simple and direct (for example, a particular flag flying over a premises means that the owners are in residence, and the flag's absence means that they are elsewhere) or complicated and voluminous (a roomful of racks of computer magnetic tape reels).

A bit is one single piece of binary information. It is either 0 or 1.

A byte is eight contiguous pieces of binary information, for example, 01001101. The most significant bit by convention is the left-handmost bit.

The least significant bit is the right-handmost bit. There are 256 (2^8) possible combinations of 0s and 1s that a byte can represent; from 00000000 to 11111111.

Octal Representation

Human handling of binary representation is time-consuming and error-prone. Octal representation is a three-binary-digit condensation technique that affords considerable convenience. If for any group of three binary digits the left-hand bit has a weighted value of four, the middle bit has a weighted value of two, and the right-hand bit has a weighted value of one, then the mapping in Table 4-1 can be seen to be true.

Any three-bit binary representation can be replaced by a single digit with a value between zero and seven. Of course, octal representation can be used in a multidigit formation. If the word "decimal" means to count in tens (our everyday system of numbers), then the decimal value of an octal number is easily calculated, (see Figure 4-1).

Hexadecimal Representation

This representation is a four-binary-digit condensation technique. It permits the value of any byte to be expressed in just two digits. The mapping methodology is in Table 4-2. The decimal value of any hexadecimal number is easily calculated (see Figure 4-2).

Binary-Coded Decimal (BCD)

This convention allows decimal numbers to be represented directly in binary-coded form. It is not efficient, because only 62% of the binary combinations possible are used. Binary-coded decimal is used in four-bit increments. See Table 4-3 for convention and example.

Teletypewriter (TTW)

This office typewriter-like machine can be connected to a computer (Figure 4-3). It has three component parts: the keyboard, the printhead, and the paper. Within the machine, there is no mechanical linkage between the keys and the printhead mechanism. The actual ink on paper printing of any character is activated in a process separate and distinct from the operator's striking the keys. A character is printed either by instruction from the computer or by striking a key. Any strike of any particular key is transmitted by code to the computer.

Table 4-1
Binary-to-Octal Mapping

$$000 = 0 \times 4 + 0 \times 2 + 0 \times 1 = 0 + 0 + 0 = 0$$
$$001 = 0 \times 4 + 0 \times 2 + 1 \times 1 = 0 + 0 + 1 = 1$$
$$010 = 0 \times 4 + 1 \times 2 + 0 \times 1 = 0 + 2 + 0 = 2$$
$$011 = 0 \times 4 + 1 \times 2 + 1 \times 1 = 0 + 2 + 1 = 3$$
$$100 = 1 \times 4 + 0 \times 2 + 0 \times 1 = 4 + 0 + 0 = 4$$
$$101 = 1 \times 4 + 0 \times 2 + 1 \times 1 = 4 + 0 + 1 = 5$$
$$110 = 1 \times 4 + 1 \times 2 + 0 \times 1 = 4 + 2 + 0 = 6$$
$$111 = 1 \times 4 + 1 \times 2 + 1 \times 1 = 4 + 2 + 1 = 7$$

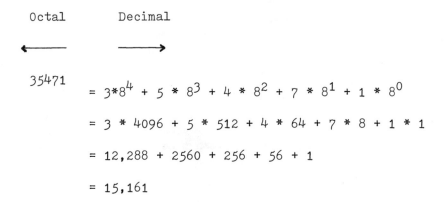

```
Octal          Decimal

←───────   ───────→

35471
        = 3*8^4 + 5 * 8^3 + 4 * 8^2 + 7 * 8^1 + 1 * 8^0

        = 3 * 4096 + 5 * 512 + 4 * 64 + 7 * 8 + 1 * 1

        = 12,288 + 2560 + 256 + 56 + 1

        = 15,161
```

Figure 4-1. An example of an octal-to-decimal conversion.

The actual character images can be solid or matrix. If they are solid, the mirror images of the character set in use reside on a drum, a ball, or a daisy wheel. If they are matrix, the coordinates of each component of each character reside in the teletypewriter memory.

Character Matrix

The displayed character consists of dots with positions in accordance with a background grid. For example, in a 5 x 7 character font in a 7 x 10 character space there are two dot columns between actual characters horizontally, and three dot rows between actual characters vertically. The user should find out whether or not the matrix font proposed for use will allow upper-case/lower-case options; whether or not the lower-case g,j,p,q, and y have descenders (tails); how the risers on b,d,f,h,k,l, and t are handled, and if underlining is possible.

Table 4-2
Binary-to-Hexadecimal Mapping

Binary Representation	Weighting	Hexadecimal Representation	Decimal Value
0000	0 + 0 + 0 + 0	0	0
0001	0 + 0 + 0 + 1	1	1
0010	0 + 0 + 2 + 0	2	2
0011	0 + 0 + 2 + 1	3	3
0100	0 + 4 + 0 + 0	4	4
0101	0 + 4 + 0 + 1	5	5
0110	0 + 4 + 2 + 0	6	6
0111	0 + 4 + 2 + 1	7	7
1000	8 + 0 + 0 + 0	8	8
1001	8 + 0 + 0 + 1	9	9
1010	8 + 0 + 2 + 0	A	10
1011	8 + 0 + 2 + 1	B	11
1100	8 + 4 + 0 + 0	C	12
1101	8 + 4 + 0 + 1	D	13
1110	8 + 4 + 2 + 0	E	14
1111	8 + 4 + 2 + 1	F	15

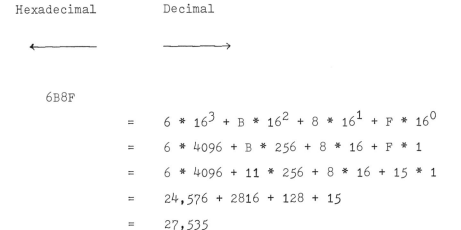

Hexadecimal $\longleftarrow$ Decimal $\longrightarrow$

6B8F

$$= \quad 6 * 16^3 + B * 16^2 + 8 * 16^1 + F * 16^0$$

$$= \quad 6 * 4096 + B * 256 + 8 * 16 + F * 1$$

$$= \quad 6 * 4096 + 11 * 256 + 8 * 16 + 15 * 1$$

$$= \quad 24,576 + 2816 + 128 + 15$$

$$= \quad 27,535$$

Figure 4-2. An example of a hexidecimal-to-decimal conversion.

Table 4-3
Binary-Coded Decimal Values

Decimal Digit	4-Bit Code
0	0000
1	0001
2	0010
3	0011
4	0100
5	0101
6	0110
7	0111
8	1000
9	1001

Example: 98 → 10011000

Since the actual dot combination of any character is contained in memory, the matrix printers can print Arabic, Farsi, Cyrillic, Katakana, and modified Roman alphabets just as well as they can print plain Roman alphabets—simply by changing the character font in the machine's memory.

Local/Remote Operation

All peripherals have two modes of operation: local and remote. Local operation means that the peripheral (a teletypewriter in this case) does not have to be connected to a computer to operate. The striking of any key will cause that character to be printed, and no communication of either event will be transmitted outside the teletypewriter. Remote operation, on the other hand, does require the device to be connected to a computer to operate. Each and every key strike is transmitted directly to the "host" computer. Character printing and carriage control are undertaken only on receipt of the requisite instructions from the "host" computer.

Full Duplex (FDX)/Half Duplex (HDX)

Remote operation of a peripheral can be undertaken in two guises. With "full duplex," a strike of a key is transmitted directly to the host computer without any immediate local action. The host computer receives the transmission and, if necessary, echoes back that same character—or perhaps echoes back some other character to the peripheral. Only upon receipt of the echo is the printing mechanism activated and the echo itself printed. Full duplex is normally a two-way, simultaneous communication. All key strikes

Figure 4-3. Several teletypewriters showing table top models with either tractor feed or roller feed and a free standing model. (Courtesy Digital Equipment Corporation.)

are received and interpreted in order. This process involves computer-generated character strings being interspersed among the keystrike echoes in correct sequence. By definition, the characters printed have been seen by the computer. Under "half duplex," when a key is struck, the character is both transmitted to the host computer and printed locally. The host computer does not echo each individual key strike. The printer still continues to print all the characters received. Generally, half-duplex is "one-way-at-a-time" communication. The host computer either receives or transmits, turn by turn, and will ignore key strikes sent during its turn to transmit. Characters can be printed that are not seen by the computer.

Control Instructions: Teletypewriter (plain)

For the automatic operation of unattended teletypewriters, various paper feed and printing mechanism control instructions are needed:

CARRIAGE RETURN	(CR)	Moves printhead to left-hand margin.
LINE FEED	(LF)	Moves paper forward one character row.
FORM FEED	(FF)	Moves printhead to left-hand margin, and advances paper to top of next sheet. (Teletypewriter must know form length and be correctly initialized.)
FORESPACE		Moves printhead one character space forward.
BACKSPACE	(BS)	Moves printhead one character space backward.
BELL	(BEL)	Sounds an audible tone in the teletypewriter.

Control Instructions: Teletypewriter (fancy)

CHANGE LINE SPACING	Line spacing is control-instruction selectable.
DOUBLE-HEIGHT CHARACTERS	Inclusion of a special control instruction in a line to be printed causes that line to be printed at twice the normal height.
UNDERLINE	Inclusion of a special control instruction in a line to be printed causes all non-blank characters in that line to be underlined.
CHANGE CHARACTER SPACING	Horizontal character spacing is control-instruction selectable.

ASCII (American Standard Code for Information Interchange)

This is a fixed-length Morse-like code. Each upper- and lower-case letter of the alphabet, each Arabic numeral, each punctuation mark, and each of several various control instructions has been assigned a unique seven-binary-digit pattern. The seven-bit code allows for the transmission of 128 characters: 32 control codes, 64 upper-case printable characters, 31 lower-case printable characters and the rub-out. Table 4-4 presents the ASCII code with octal and hexadecimal equivalents.

The individual control codes do not have universal application among computer and peripheral manufacturers. Neither have the manufacturers adopted a standard word length for transmission. It is possible to have the seven-bit code with or without an eighth bit (set to zero or to one) and with or without a parity bit (set to even parity or to odd parity). When actually transmitted, this seven-, eight-, or nine-bit data code will be preceded by one start bit (set to zero) and followed by one, possibly two, stop bits (set to one). Figure 4-4 outlines the variations possible. Unless the transmitter and the receiver are set to the same character and transmission formats, the data transmitted will be unintelligible.

Serial Communications (EIA-RS232C/20mACL)

There are two methods of transmitting the ASCII information in serial form: RS232C or 20 milliampere current-loop. On the EIA-type 25-pin connector, the following pin assignments are made:

	Pin #	
RS232	1	Frame ground
	2	Transmit data
	3	Receive data
	7	Signal ground
20mACL	17 & 24	Transmitter
	23 & 25	Receiver

RS232C is normally used for transmission involving modems and telephone lines. 20mACL is ordinarily used for transmissions where the peripheral is coupled directly to the computer. In serial transmission the least significant bit is transmitted first (the bit that distinguishes a *C* from a *B*) and the most significant bit is transmitted last (the bit that distinguishes an *r* from a *2*). See Table 4-4.

In actual transmission of character strings there is no deliberate pause between characters. If the receiver is set to receive 10-bit transmission format containing an eight-bit data format, the receiver will read in 10 bits in sequence at the specified baud rate. To affirm that it is indeed a valid 10-bit character sequence, the receiver checks that bit #1 is a 0 and bit #10 is a 1. If they are both correct, the middle eight bits are read in as a valid data string. If either the first bit is a 1 or the last bit is a 0, the transmission

Table 4-4
The 128-Member ASCII Code

BINARY	OCT	HX	CTRL			BINARY	OCT	HX		BINARY	OCT	HX		BINARY	OCT	HX	
0000000	000	00	@	NUL	null	0100000	040	20	blank	1000000	100	40	@	1100000	140	60	`
0000001	001	01	A	SOH	start of header	0100001	041	21	!	1000001	101	41	A	1100001	141	61	a
0000010	002	02	B	STX	start of text	0100010	042	22	"	1000010	102	42	B	1100010	142	62	b
0000011	003	03	C	ETX	end of text	0100011	043	23	#	1000011	103	43	C	1100011	143	63	c
0000100	004	04	D	EOT	end of transmission	0100100	044	24	$	1000100	104	44	D	1100100	144	64	d
0000101	005	05	E	ENQ	enquiry	0100101	045	25	%	1000101	105	45	E	1100101	145	65	e
0000110	006	06	F	ACK	positive acknowledge	0100110	046	26	&	1000110	106	46	F	1100110	146	66	f
0000111	007	07	G	BEL	ring bell	0100111	047	27	'	1000111	107	47	G	1100111	147	67	g
0001000	010	08	H	BS	backspace	0101000	050	28	(	1001000	110	48	H	1101000	150	68	h
0001001	011	09	I	HT	horizontal tabulation	0101001	051	29	)	1001001	111	49	I	1101001	151	69	i
0001010	012	0A	J	LF	line feed	0101010	052	2A	*	1001010	112	4A	J	1101010	152	6A	j
0001011	013	0B	K	VT	vertical tabulation	0101011	053	2B	+	1001011	113	4B	K	1101011	153	6B	k
0001100	014	0C	L	FF	form feed	0101100	054	2C	,	1001100	114	4C	L	1101100	154	6C	l
0001101	015	0D	M	CR	carriage return	0101101	055	2D	-	1001101	115	4D	M	1101101	155	6D	m
0001110	016	0E	N	SO	shift out	0101110	056	2E	.	1001110	116	4E	N	1101110	156	6E	n
0001111	017	0F	O	SI	shift in	0101111	057	2F	/	1001111	117	4F	O	1101111	157	6F	o
0010000	020	10	P	DLE	data link escape	0110000	060	30	0	1010000	120	50	P	1110000	160	70	p
0010001	021	11	Q	DC1	device control 1, x-on	0110001	061	31	1	1010001	121	51	Q	1110001	161	71	q

Binary	Octal	Hex	Char	Mnemonic	Name
0010010	022	12	R	DC2	device control 2, aux-on
0010011	023	13	S	DC3	device control 3, x-off
0010100	024	14	T	DC4	device control 4, aux-off
0010101	025	15	U	NAK	negative acknowledge
0010110	026	16	V	SYN	data synchronization
0010111	027	17	W	ETB	end of text block
0011000	030	18	X	CAN	cancel
0011001	031	19	Y	EM	end of medium
0011010	032	1A	Z	SUB	substitute
0011011	033	1B	[	ESC	escape
0011100	034	1C	\	FS	file separator
0011101	035	1D	]	GS	group separator
0011110	036	1E	^	RS	record separator
0011111	037	1F	_	US	unit separator

Binary	Octal	Hex	Char
0110010	062	32	2
0110011	063	33	3
0110100	064	34	4
0110101	065	35	5
0110110	066	36	6
0110111	067	37	7
0111000	070	38	8
0111001	071	39	9
0111010	072	3A	:
0111011	073	3B	;
0111100	074	3C	<
0111101	075	3D	=
0111110	076	3E	>
0111111	077	3F	?

Binary	Octal	Hex	Char	Name
1010010	122	52	R	
1010011	123	53	S	
1010100	124	54	T	
1010101	125	55	U	
1010110	126	56	V	
1010111	127	57	W	
1011000	130	58	X	
1011001	131	59	Y	
1011010	132	5A	Z	
1011011	133	5B	[	
1011100	134	5C	\	
1011101	135	5D	]	
1011110	136	5E	^	
1011111	137	5F	_	underline

Binary	Octal	Hex	Char	Name
1110010	162	72	r	
1110011	163	73	s	
1110100	164	74	t	
1110101	165	75	u	
1110110	166	76	v	
1110111	167	77	w	
1111000	170	78	x	
1111001	171	79	y	
1111010	172	7A	z	
1111011	173	7B	{	
1111100	174	7C	|	
1111101	175	7D	}	
1111110	176	7E	~	tilde
1111111	177	7F		rub-out

start bit, 7-bit code, 1 stop bit

start bit, 7-bit code, 2 stop bits

start bit, 7-bit code, odd parity bit, 1 stop bit

start bit, 7-bit code, odd parity bit, 2 stop bits

start bit, 7-bit code, even parity bit, 1 stop bit

start bit, 7-bit code, even parity bit, 2 stop bits

start bit, 7-bit code, 8th bit 0, 1 stop bit

start bit, 7-bit code, 8th bit 0, 2 stop bits

start bit, 7-bit code, 8th bit 1, 1 stop bit

start bit, 7-bit code, 8th bit 1, 2 stop bits

start bit, 7-bit code, 8th bit 0, odd parity bit, 1 stop bit

start bit, 7-bit code, 8th bit 0, odd parity bit, 2 stop bits

start bit, 7-bit code, 8th bit 0, even parity bit, 1 stop bit

start bit, 7-bit code, 8th bit 0, even parity bit, 2 stop bits

start bit, 7-bit code, 8th bit 1, odd parity bit, 1 stop bit

start bit, 7-bit code, 8th bit 1, odd parity bit, 2 stop bits

start bit, 7-bit code, 8th bit 1, even parity bit, 1 stop bit

start bit, 7-bit code, 8th bit 1, even parity bit, 2 stop bits

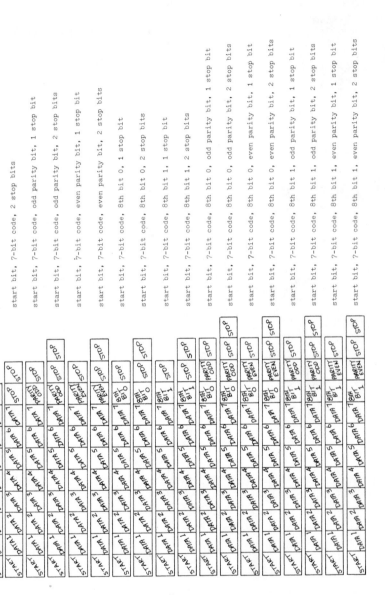

Figure 4-4. Character bit patterns in serial communications.

sequence has been misframed. The bits will be shuffled one bit to the left, the old #1 bit dropped, and a new #10 bit read in. The validation procedure is then repeated. Should the receiver at any time be out of sequence with the transmitter, the receiver will self-correct as soon as there is even a momentary pause between transmissions of characters. The pause serves as a gate to make the framing of the subsequent characters correct. The pause has been read in, at the baud rate, as a string of 1s.

Baud Rate

The baud rate is the number of bits transmitted per second in serial communications. The baud rate is sometimes written bps. If the data transmitted contains 10 bits per character, then a rate of 9600 baud will represent a rate of 960 characters per second.

The total elapsed times at different baud rates for transmission/building of various complete video and printed displays are presented in Table 4-5 and Table 4-6, respectively.

The usual baud rates are 75, 110, 150, 300, 600, 1200, 1800, 2400, 4800, 9600, 19200, and 38400. User choice is switch-selectable at the peripheral and wire-wrap selectable at the computer.

Parity is an extra bit included in a data format to ensure data integrity. It is set either to "even" or to "odd." If even parity is chosen, then in each and every data format transmitted the number of bits set to 1 is made an even number by setting or not setting the parity bit itself to 1. In a like fashion, if odd parity is chosen, then in each and every data format transmitted the number of bits set to 1 is made an odd number by setting or not setting the parity bit itself to 1. If on receiving data in odd parity an even number of 1s is counted in the data format, then a 1 bit has been dropped or a 1 bit has been picked up and the character is invalid. For *even* parity transmissions, *odd* 1 counts invalidate the data.

Parallel Communications

The eight bits of the data code can also be transmitted between the peripheral and the computer in parallel fashion over eight parallel conductors. Another two conductors are required: one conductor to signal that data is assembled and ready to be read, and one conductor to signal that it has been read. For two-way communication, 20 or more conductors are required. Parallel communications are very fast but only satisfactory over very short distances (intracomputer room).

RS422

This is RS232 for long distances.

Table 4-5
CRT Display Build Times in Seconds

Baud Rate	Display 64 col × 20 rows	80 col × 24 rows	80 col × 48 rows
2400	5.3	8.0	16.0
4800	2.7	4.0	8.0
9600	1.3	2.0	4.0
19200	0.7	1.0	2.0
38400	0.3	0.5	1.0

Table 4-6
Page Printing Time

LPM	8½ × 11 72 col × 66 rows	11 × 15 132 col × 66 rows
120	33 secs	33 secs
300	13. secs	13 secs
600	7 secs	7 secs
BAUD		
110	7 min, 55 secs	14 min, 31 secs
300	2 min, 38 secs	4 min, 50 secs
1200	40 secs	1 min, 13 secs
4800	10 secs	18 secs
9600	5 secs	9 secs

The Video Display Terminal

The bulk of the input and output to and from a minicomputer system is by a video display terminal, commonly called a CRT (cathode ray tube). A CRT has three component parts: the video monitor, the video generator, and the keyboard (Figure 4-5).

The video monitor resembles a television but has a much higher performance. The tube itself is a large, empty glass bottle with a phosphor on the inside surface of the flattened end. The diagonal measurement of the picture can range from five inches to 25 inches. If the phosphor is struck by a beam of high-velocity electrons, it will glow momentarily. A picture in glowing phosphor is created by sweeping the screen from top to bottom (usually in 264 non-interlaced lines) with an electron beam 60 times a second, varying the beam's strength so that there are alternate light and dark sections. This is a scan rate of 15,840 Hertz. If the screen has 24 rows of characters, and

Figure 4-5. A black-and-white CRT showing the screen and keyboard. (Courtesy Lear Siegler, Inc.)

each character is a 5 x 7-dot matrix; then the 264 lines comprise 14 blank lines at the top, 24 sets of seven lines (the character rows), 23 interleaved sets of three blank lines (the spaces between character rows), and 13 blank lines at the bottom. If the row is 80 characters wide and each five-dot-wide character is in a seven-dot cell, then there are 560 possible dot spaces (or pixels) in a line. In one second the beam has targeted 8,870,400 dot positions. As the beam actually zig-zags down the screen, creating dots left to right and being blanked right to left, the dot generation rate is in excess of 17 megaHertz; to dot or not to dot every 60 nanoseconds.

This is the function of a video generator. CRTs accept the ASCII-coded characters (already mentioned) by RS232 or 20 mACL interconnections and store them in an inner memory. Such CRT display memories come in many sizes:

12 rows × 32 columns
20 rows × 64 columns
24 rows × 80 columns
24 rows × 132 columns
25 rows × 80 columns
48 rows × 80 columns
60 rows × 80 columns

The video generator then works through the stored characters dot line by dot line: issuing the dots, the horizontal sync signals for each dot row, and the vertical sync signals for each frame to the video tube. The size of the tube has no effect (other than legibility) on the number of characters in a display.

A character accent is possible. Reverse video (black dots in a white background), a dim (50% or less of normal), and blinking (usually 80% on, 20% off) are the usual attributes.

The CRT keyboard (Figure 4-6) is much like a teletypewriter keyboard, except that it carries some unusual keys. To indicate on the screen exactly where the next character will be entered, either an underline or a solid-filled character is blinked. This is the cursor. The unusual keys (↑, ↓, →, ←) are cursor control keys: cursor up one line, cursor down one line, cursor forespace, and cursor backspace. "Home" will cause the cursor to go to the top-left corner or the bottom-left corner, depending on the particular make and model of the CRT. "Erase" or "blank" causes the screen to go blank. Cursor control is when the horizontal and vertical coordinates of the desired position of the cursor are given to the video generator, rather than giving repeated instructions to move to the adjacent cell.

Ways to Get Video Displays Out of a Computer (Figure 4-7)

1. The easiest and most expensive method is to use a single-channel interface card wired with a four-conductor cable to a complete CRT that includes keyboard, video generator, and screen.

2. The cheapest way to get a second display is to use a "slave" monitor. Take a second composite video output from the video generator in Method 1, and run a coaxial cable to a second screen. The second display has no keyboard capacity and shows only what is on the master screen.

Figure 4-6. The layout of a teletypewriter or CRT keyboard.

3. Put the video display on a normal television. It is convenient, but the picture quality is poor. Take the video signal in Method 2, and connect it to a radio frequency converter for connection to the antenna lugs of a television.

4. Multiple "slaves" can be used with good effect. Take the video signal from Method 1, and amplify it enough in a video amplifier to power several slave monitors.

5. Several different video displays can be taken off of one interface channel. Wire the interface to a string of successive software-pollable CRTs. The actual CRT selected at any given time is imbedded in the ASCII string. At any one time it is possible to sequence up to 254 different displays to 254 different CRTs through one interface channel. One or several keyboards may be allowed in the string.

6. Several complete CRTs can be daisy-chained together on one four-conductor cable. They all will show the same display, and the computer cannot differentiate between keyboards. This method is expensive because of the redundant use of video generators and keyboards. Method 4 is a cheaper alternative.

7. The most effective way is to use a multichannel (MUX) interface with several complete CRTs. The MUX replaces several single-channel interface cards. The congestion on the bus of the computer is reduced. The programming is easier because one single device code replaces several device codes. The overhead on the interface handling within the operating system is much reduced.

Video Graphics

The video generator is enhanced enough so that the pixels in the display can be turned on and off independently of each other. This allows lines, rectangles, circles, and free-form characters to be displayed. The video gen-

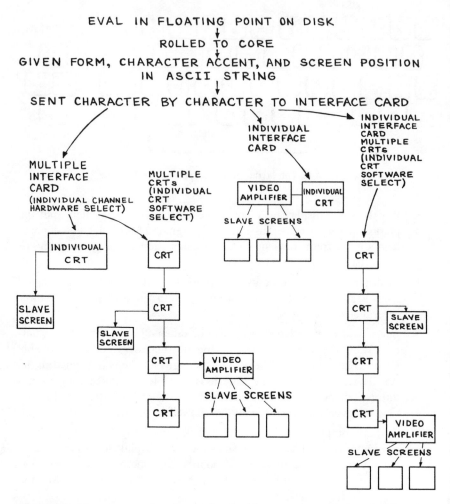

Figure 4-7. The process and equipment involved in generating a display of a number (in humanly understandable form) from a stored value (in floating-point notation).

erator will allow either single pixels to be turned on/off or will turn on/off the pixels in a straight line between two given pairs of coordinates. This necessitates a sequence of instructions until a complete picture is built.

RS-170

The RS-170 is a black-and-white composite video signal that incorporates horizontal and vertical sync. It is used to drive monitors.

The Front-End

The front-end is the chemical process's interface with the computer (Figures 4-8, 4-9). It takes in the real world's analog measurements and off/on signals and presents them to the computer in machine-digestible form. It also takes the machine's digital output instructions and presents them to the real world as either analog values or on/off signals. The front-end has two parts: the analog-to-digital converter (AI) for the analog measurements and the digital input/digital output converter (DIDO) for the on/off switching and analog outputs.

The AI

This has a scan rate which is a function of the type of inputs used. For a system of all high-level signals (>1.25V), the AI can scan at a rate of 10,000 points per second for a maximum of 128 signals. Any one channel can be read repeatedly only at a lower rate of 200 times per second. For a system of millivolt signals, the AI scan rate must be reduced to 20 points a second, with a maximum of 64 signals. Some AIs are available with wide-range capability which handle both high-level and low-level signals. Such systems typically scan at 200 samples per second. The number of signals connected can be a maximum of 512, 1024, or 2,048, depending on the manufacturer. The manufacturer also determines whether the inputs are mapped on to 11-bit scales (-2048 to $+2047$), 12-bit scales (-4096 to $+4095$), 13-bit scales (-8192 to $+8191$), or 16-bit scales ($-65,536$ to $+65,535$). The connection of field wiring to the computer is through racks of screw terminals (Figure 4-10). Table 4-7 gives the range of input measurements that can be handled, but the entire range is not handled on the typical system. It is possible within limits to mix and match a great variety of incoming signals through one AI unit. A schematic of the complete flow of information from its source (the real-world variable) to its destination (storage on disk in floating-point notation) is shown in Figure 4-11.

There is some incompatability when a 0V-5V or 1V-5V controller output, representing a 1000-increment scale, is mapped over a 0V-5V computer input representing a 2047-integer scale. For the former 0V-5V range, every real-world 20th increment will be stored as the 21st in the computer. For the latter 1V-5V range, the stairstep effect is even worse, since there are 1638 integer values for the 1000 positions. For example:

1. An integer value of one represents a pulse position of 0.6
2. An integer value of two represents a pulse position of 1.2
3. An integer value of three represents a pulse position of 1.83

Figure 4-8. A double-rack "front-end" showing a few of the signal conditioning cards and controller cards already inserted in the rack. (Courtesy Computer Products, Inc.)

4.   An integer value of four represents a pulse position of 2.44

5.   An integer value of five represents a pulse position of 3.1.

The series becomes one, one, two, two, and three. It is possible not to register real-world movements of one increment.

The mismatch between the controller 1000 scale and the computer's integer scales can be avoided. For 0V-5V signals, a computer input range of 5.12 volts allows a precise match of increment scale and integer scale. Better yet is an analog range of 1023 increments full scale. For 1V-5V analog outputs representing 1000-increment scales, the linearity is so mismatched that the output range will have to be rearranged to 1V-3.5V if a 1023-increment scale is to be retained. This modification is virtually impossible to accomplish. An easier modification is a 1V-4.75V range representing a 512-increment scale.

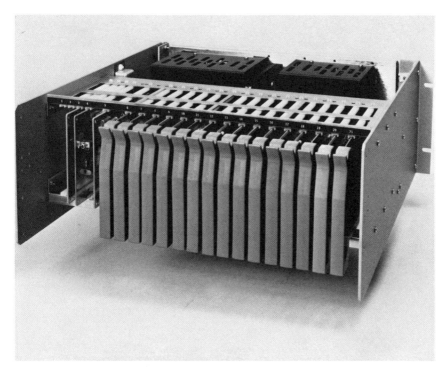

Figure 4-9. A single-rack "front-end" showing the cable termination edge connectors which attach the cables from the screw terminals to a full complement of signal conditioning cards. (Courtesy Computer Products, Inc.)

The DIDO

For an incoming on/off switch (24V or 48V), the DIDO senses the change. It releases the information to the processor either immediately with an interrupt or later under interrogation. An outgoing on/off switch is set until it is reset (a flip-flop), or it is set for a certain trimpot-adjustable time or programmable duration from as short as 100 microseconds to as long as 18 minutes (a single-shot). The possible analog output ranges for a 0-1023 10-bit scale are shown in Table 4-8. Table 4-9 summarizes scales of square root flow and linear percent against input analog voltage scales of 0V-5V and 1V-5V.

DIDOs vary in size. One manufacturer provides a DIDO that sets or senses from one to 256 banks of eight switches. Another manufacturer offers a DIDO that handles from one to 16 banks of 16 switches.

Various additional functions can be found on DIDOs: time-countdown cards, input pulse-counting cards, frequency-counting cards, pulse-train generating cards, and time-of-year cards with self-contained battery backup.

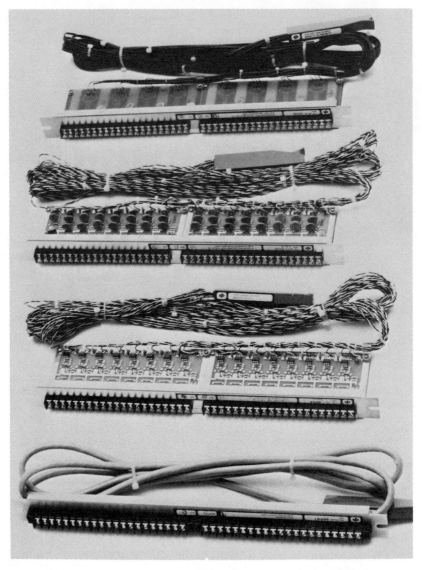

Figure 4-10. Various screw terminal strip assemblies. (Courtesy Computer Products, Inc.)

Table 4-7
Input Analog Measurement Ranges

Volts		Millivolts	
Minimum	Maximum	Minimum	Maximum
−2.0V	100.0V	−1024mV	1024mV
−2.0V	50.0V	−500mV	500mV
−10.24V	10.24V	−200mV	500mV
0.0V	10.24V	−200mV	200mV
−2.0V	10.0V	−102.4mV	102.4mV
−200mV	10.0V	−100mV	100mV
−5.12V	5.12V	−50mV	50mV
0.0V	5.12V	−10.24mV	10.24mV
−2.0V	5.0V	−10mV	10mV
−200mV	5.0V		
−2.56V	2.56V	Milliamperes	
0.0V	2.56V	Minimum	Maximum
−2.0V	2.0V	−20mA	50mA
−1.28V	1.28V	−20mA	20mA
0.0V	1.28V	−10mA	10mA
−1.024V	1.024V	−5mA	5mA
−1.0V	−1.0V	−1mA	1mA
−200mV	1.0V		

The System

The way information is presented to and displayed by a computer has been discussed. Now the two main features of the computer proper will be described: memory and disk.

Memory

The memory is the workspace of the computer (Figures 4-12 and 4-13). The larger it is, the more powerful the machine. Large memories allow either one large program or several short programs to reside wholly within the machine. The word length of the machine sets an upper limit on the size of the memory it can handle. Memory extension devices or memory management can be used at extra cost to increase the physical size of the memory, although at any one instance the processor still can access only the original maximum memory size as set by the word length.

Controllers

These are the devices that coordinate the transfer of data between the computer memory and the rotating disk. When they are informed of the

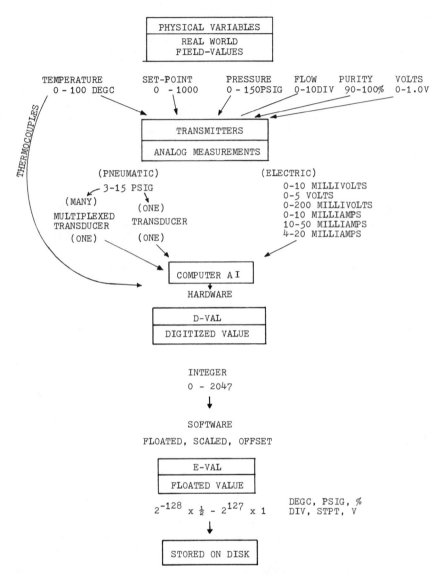

Figure 4-11. The process involved in creating stored values (in floating-point notation) from real-world physical variables.

Table 4-8
Output Analog Ranges

Volts	
Minimum	Maximum
−10.24V	10.24V
0.0V	10.24V
0.0V	10.0V
−5.12V	5.12V
0.0V	5.12V
1.0V	5.0V

Milliamperes	
Minimum	Maximum
4mA	20mA
10mA	50mA

memory address, the disk address (drive, surface, track, sector, and word), and the number of words to be transferred, they take over the operation and signal either when the data transfer process is complete or when an error has occurred. Controllers can handle more than one drive. It is unusual for a system to have more than one controller.

Disks (Figures 4-14, 4-15, and 4-16)

Minicomputers and microprocessors use three kinds of disks: fixed-head disks, moving-head disks, and floppy disks. Disks are revolving magnetic media on which are arranged multiple circular (not spiral) tracks. Each track is divided into sectors, and each sector is divided further into bytes. Data can either be read from the disks or written to them. The smaller, older fixed-head disks allowed records of varying lengths (from one word to 2048 words) to be transferred. The newer, larger moving-head disks still allow one-word transfers, although it is more likely that only complete sectors (128 or 256 bytes) can be transferred.

In terms of computer execution speeds disks are inexorably slow. To read or write any specific sector, that sector must be under the read/write head. It may mean a delay of one disk revolution. In a moving-head disk it may mean a delay while the head travels from the first track to the last track.

Fixed-head disks.   With this type of disk, there is a read/write head in a fixed position for each track. Logically adjacent records will not be phys-ically adjacent on the disk. They will be alternated either every second or

Table 4-9
**Input Voltages and Corresponding Linear Percent
and Flow Division Values**

Volts	11-Bit DVAL	0-5 Volt Analog Input Flow Divisions	%	1-5 Volt Analog Input Flow Divisions	%
0.00	0	0.00	0		
0.05	20	1.00	1		
0.20	82	2.00	4		
0.45	184	3.00	9		
0.50	205	3.16	10		
0.80	328	4.00	16		
1.00	409	4.47	20	0.00	0
1.04	426			1.00	1
1.16	475			2.00	4
1.25	512	5.00	25		
1.36	557			3.00	9
1.40	573			3.16	10
1.50	614	5.48	30		
1.64	671			4.00	16
1.80	737	6.00	36	4.47	20
2.00	819	6.32	40	5.00	25
2.20	901			5.48	30
2.44	999			6.00	36
2.45	1003	7.00	49		
2.50	1023	7.07	50		
2.60	1064			6.32	40
2.96	1212			7.00	49
3.00	1228	7.75	60	7.07	50
3.20	1310	8.00	64		
3.40	1392			7.75	60
3.50	1433	8.37	70		
3.56	1457			8.00	64
3.75	1535	8.66	75		
3.80	1556			8.37	70
4.00	1638	8.94	80	8.66	75
4.05	1658	9.00	81		
4.20	1719			8.94	80
4.24	1736			9.00	81
4.50	1842	9.49	90		
4.60	1883			9.49	90
5.00	2047	10.00	100	10.00	100

Figure 4-12a. The inner detail of a 4K dynamic RAM chip. (Courtesy Mostek.)

Figure 4-12b. The 4K dynamic RAM chip mounted in an assembly package (Courtesy Mostek.)

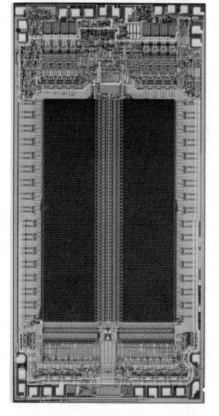

Figure 4-12c. The inner detail of a 16K dynamic RAM chip. (Courtesy Mostek.)

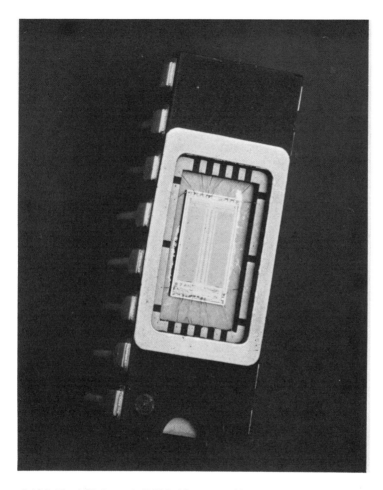

Figure 4-12d. The 16K dynamic RAM chip mounted in an assembly package. (Courtesy Mostek.)

every fourth physical record. Logically adjacent records one, two, and three may occupy the first, third, and fifth or first, fifth, and ninth physical positions in a track. Consequently, a transfer of 2048 records may take two or four revolutions of the disk to complete, provided the disk can register its timing mark each revolution. At 1800 rpm, one revolution takes 33.3 milliseconds. A transfer of 2048 records may take up to 167 milliseconds (five revolutions) from start to finish if recorded in every fourth position. For a transfer, that is a long time.

Great care has to be taken in sequencing information from a disk. Access of adjacent records, one at a time, can be no faster than one per revolution or one per 33.3 milliseconds. If the operation is to read a record, change it, write it back, and repeat the process on the next word, the operation slows

Figure 4-13. A printed circuit board carrying 128K of 12-bit MOS memory, consisting of 96 16K dynamic RAM chips similar to those in Figure 4-12d. (Courtesy Plessey Peripheral Systems.)

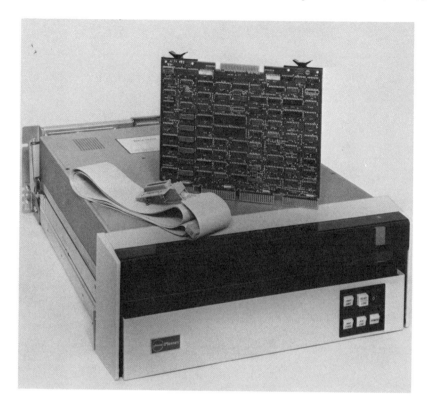

Figure 4-14. A hard-disk, bulk memory system and controller board. (Courtesy Plessey Peripheral Systems.)

to one complete cycle every 66.7 milliseconds, presuming, of course, that the changing takes less than 33.3 milliseconds. It will be quicker to read the whole track, do the 2048 changes, and write the whole track back. The reading and writing will then take 333 milliseconds (10 revolutions) at worst compared to 136,602 milliseconds (4096 revolutions; more than two minutes) reading and writing individual records. In fact, by careful timing and comparison of the record-transfer counter, the changing can be made simultaneous with the reading and writing of the whole track, reducing the whole operation to a minimum time. The changing begins when the first record appears in memory and is completed just before the last record is written back to the disk. Unless the positions of records on a disk are optimized, or the computer software has been written to accommodate multilevel processing (i.e., during a disk transfer, progress is made in some other program already in memory), frequent disk accesses will degrade a system's overall perfor-

Figure 4-15. A 300-megabyte multiplatter moving-head, bulk memory disk drive. (Courtesy Plessey Peripheral Systems.)

Figure 4-16. A double-drive, floppy-disk, bulk memory system. (Courtesy Data Systems Design, Inc.)

mance. The access of adjacent records will mean that the worst-case access times are the norm. Do not use average access times in calculating the program's overall computation times.

A typical platter holds 128 tracks. A system can have up to four platters for an overall word storage of 1,048,576 words.

Moving-head disks. These disks have one read/write head for each surface that is moved track to track. The development of moving-head technology was accompanied by improvements in disk-recording densities and recording methods. Adjacent logical records occupy adjacent physical positions. However, there is still rotational latency to which is added the extra latency of track-to-track seek times. The comments made previously about accessing fixed-head disks apply as well to moving-head disks.

Although one-word transfers may be technically feasible, software packages will agglomerate four or eight sectors into blocks and only allow block transfers. This means that for one word to be changed, one block (possibly 2048 words long) must be read into memory, and then restored to the disk. Table 4-10 presents various typical disk characteristics.

Floppy disks. These are small, flexible magnetic media used to transfer information in bulk to and from systems. Sometimes the smallest of systems will use floppy disks as the bulk storage medium. Ordinarily, a floppy will have 77 tracks of 26 sectors of 128 bytes each for a total of 2002 tracks and 256,256 bytes. Only a complete sector at a time can be accessed. It is not possible to transfer single bytes. Some floppy-disk controllers will allow up to 30 sectors per track and alternatively fewer sectors of greater length. But only a track of 26 sectors of 128 bytes is the recognized industry standard. Double-density recording permits 256 bytes per sector and 512,512 bytes per disk. Double-sided recording allows both sides of the floppy to be used. Floppy-disk controllers control two or four disks each. A minicomputer system can have as many as eight controllers for a total floppy storage of 32 million bytes.

Floppies revolve slowly: 360 rpm, or 167 milliseconds per revolution. It is not possible to transfer the 128 or 256 bytes between disk controller and memory fast enough to permit access to two adjacent sectors. In fact, the actual transfer process may take more than six milliseconds, and even alternate sectors cannot be addressed. Six milliseconds is the time it takes a sector to pass a read/write head.

Transfer from the disk to the buffer and from the buffer to the memory has to be less than 23 microseconds per byte of a 256-byte sector to get the best possible access out of a floppy. If the physical sectors are logically numbered in sequence (1, 2, 3. . .25, 26), it must take 26 revolutions of the disk to read the entire track. If the physical sectors are reformatted to read

Table 4-10
Typical Moving-Head Disk Configurations

Drives	Surfaces Per Drive	Tracks Per Surface	Sectors Per Track	Bytes Per Sector	Bytes One Drive	Bytes Total System	RPM	Milli-secs 1 Rev.	1 Track Seek Milli-secs	Max Seek Milli-secs	Min Transfer Words	Max Transfer Words
A 4	2	256	40	256	5,242,880	20,971,520	2400	25.0	15.0	100	1	5,120
B 1	5	823	32	512	67,420,160	67,420,160	3600	16.6	6.0	55	1	65,535
C 8	2	200	12	512	2,457,600	19,660,800	1500	40.0	10.0	85	1	65,536
D 8	3	411	20	512	12,625,920	101,007,360	2400	25.0	8.0	75	1	65,536
E 8	3	815	22	512	27,540,480	220,323,840	2400	25.0	6.5	71	1	65,536
Winchester Disks (256-Byte Transfers)												
14″	2	404	60	256	12,410,880		2964	20.2	20	130		
14″	4	404	60	256	24,821,760		2964	20.2	20	130		
8″	2	256	32	256	4,194,304		3125	19.2	19	140		
8″	4	256	32	256	8,388,608		3125	19.2	19	140		

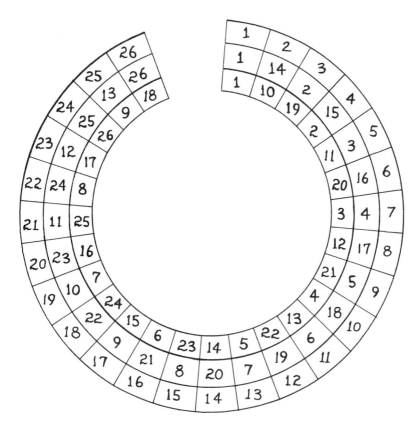

Figure 4-17. Various alternate sector formats for a floppy disk.(Courtesy Data Systems Design, Inc.)

logically in physical sequence (1, 14, 2, 15...13, 26), it may be possible to access alternate sectors and read or write the entire track in two disk revolutions (see Figure 4-17). A floppy cannot be accessed any faster. It may be necessary to reformat to every third or every fourth physical sector, which will increase the times to three and four disk revolutions, respectively; but this is still a considerable saving over 26 revolutions. Unless the sectors are offset track to track, an attempted spiral write of a floppy will take three revolutions per track in the alternating sector format. The track-to-track seek time requires a two- or three-sector offset to prevent missing that first sector at track-stepping time. The sectors in a track need not be numbered one through 26, and the sectors need not be in sequence. All that is required is that they be 26 different numbers between one and 255. Unusual sector numbers can impart some confidentiality to a floppy's data.

Figure 4-18. Unmounted Winchester disks: 14 inches left, 8 inches middle, and 5.25 inches right. (Courtesy Shugart Associates.)

Winchester disks (Figure 4-18).   This technology encloses a multiplatter disk and its read/write heads in a sealed enclosure. It allows fantastic recording densities at economical prices. As with floppies, only complete sectors can be transferred. In fact, there is enough compatibility between the smaller Winchester disk drives and floppy-disk drives that if the software will support it, a floppy of 256,256-byte storage can be replaced physically in the same slot by a disk of 8,388,608-byte storage capacity. The larger Winchester technology disks carry 24.8 megabytes of storage in 5¼ inches of standard 19-inch rack.

The Arithmetical Logical Unit (ALU)

This is the heart of the machine where all the logical decisions and data manipulations are made (Figure 4-19). In sequencing through a program in

Figure 4-19a. The inner detail of a Z-80 microprocessor CPU chip. The black stripes around the edge are the electrical conductors between the chips and the legs of the mounting package. (Courtesy Mostek.)

memory the word stored at the address contained in the "program-location" counter is read into the ALU, and the program location is incremented. The word is decoded and acted upon. If the decoded instruction requires that data from a location in memory be loaded into a register, or if a register is to be written to a location in memory, it is done. If data in different registers are to be arithmetically manipulated, the arithmetic is accomplished. If by the testing of a register the sequence of the program is altered, the beginning address of the new sequence is loaded into the program-location counter. If the word actually instructs that the program sequence be diverted to a new location (a jump), the address of that new location is loaded into the program-

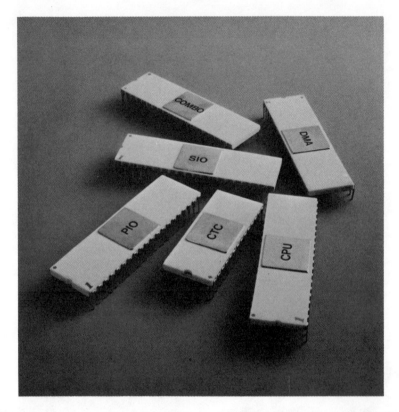

Figure 4-19b. An assortment of LSI packages, including the CPU chip detailed in Figure 4-19a. There are also the parallel input output (PIO), the serial input output (SIO), a controller for direct memory access (DMA), a combination package (COMBO), and a multi-channel clock timer counter (CTC). (Courtesy Mostek.)

location counter. At the completion of the total instruction cycle, the word stored at the address contained in the program-location counter is read, and the cycle is repeated once more.

Configuring a System

How is a minicomputer system assembled? What decisions and choices are made in building up the hardware available into a complete, operable entity? The answer is "many, many, coordinated decisions," for it is a process of selection that may be repeated several times for several different configurations. In this field of endeavor there are many different ways of achieving the targeted results. It is quite probable that there will be no clear-cut "best way." In fact, it is more than likely that tomorrow's product release will make today's second choice the first choice.

A computer system that is to be used for process control in real time with human operator interface (and is to maintain a management data base), will be a large, complicated installation, and quite definitely the most demanding use to which a computer can be put. It is most important in a system of this sort that the ultimate expectations are known at the outset so that the initial equipment will be bought with sufficient power and rack space to support all the intended expansions and peripherals. How is such a system built? How is it expanded from the initial minimum configuration to a supersophisticated bus banger? The steps themselves are quite logical and are pictured in Figures 4-20a through 4-20g.

Some users prefer to get a complete system from one manufacturer. It simplifies purchasing, installation, and future maintenance. Other users prefer to buy particular peripherals from particular manufacturers for financial or performance reasons, building their own systems from individual components. Whichever way is chosen, each purchasing decision involves weighing one factor against another: purchase price, delivery time, maintenance cost, technical support documentation, compatibility, suitability, expandability, and experience—yours, theirs, and other people's.

The decisions begin with what alternatives exist for the CPU. This is the keystone around which all the expansions and peripherals will be fitted. It is necessary to make a wise and careful selection. Word size, instruction execution speed, available software, peripheral availability, mechanical quality, electronic quality, instruction set, and bus throughput are the ultimate performance parameters—not only of the immediate process but of the ultimate system. Do not make a precipitous decision about the CPU (Figure 4-20a).

How large a memory is required? How fast should it be? Should it be core, read-only MOS, random-access MOS, or bipolar? The larger memories enable larger programs to be run without chaining, or several smaller programs to be concurrent in core; or they allow a memory-based operating system. The smaller memories must chain large programs and are very restrictive of operating systems of any kind. It is necessary for the CPU and memory to be connected by a backplane. Generally, the choice of the CPU will determine automatically what kind of backplane is used, but the user must decide on the physical size of the backplane. A small one with slots (two or four) sufficient for the CPU, memory, and one or two peripheral interfaces; or a large backplane (73 or 144 slots) big enough to receive all the intended expansion (Figures 4-20b and 4-21).

One or several power supplies will be required to supply power to the backplane. The sizing will be determined largely by the ultimate size of the system. A processor console on the system may be a very helpful feature to

Figure 4-20a. The progress of building a hardware system from its individual components: CPU alone.

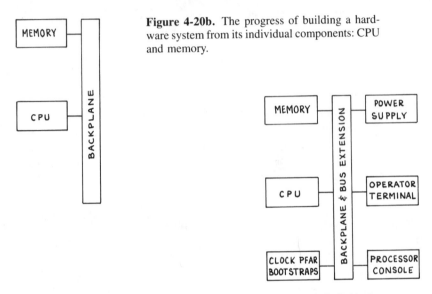

Figure 4-20b. The progress of building a hardware system from its individual components: CPU and memory.

Figure 4-20c. The progress of building a hardware system from its individual components: a minimal configuration (limited program editing).

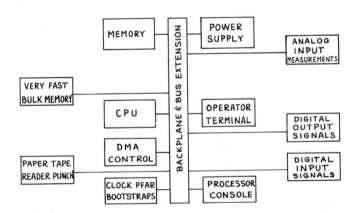

Figure 4-20d. The progress of building a hardware system from its individual components: a minimal process-control configuration.

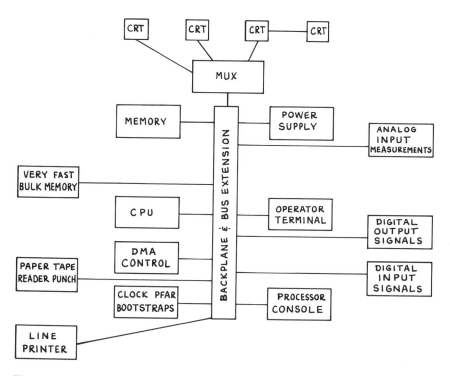

Figure 4-20e. The progress of building a hardware system from its individual components: a process-control configuration.

add. This is a bank of lights and a rotary switch selector plus key switches or a key pad (Figure 4-23). The lights are used to indicate the states of the processor's inner workings. Individual combinations of the various states are selected by different positions of the rotary switch. The key switches are used to enter addresses or data whenever there are either hardware or software problems. The operator terminal is a teletypewriter through which ultimate control of the computer is exercised by the operator. It is used to enter commands and record messages. The processor will need a clock to keep track of the time and a power-fail auto-restart device for power outages. A power-fail auto-restart (PFAR) halts the processor as the power fails and restarts the processor when the power is restored. Restarting the processor will require one of various bootstraps. These are short, dedicated programs temporarily read into memory which allow the computer to cold start from a system disk, a paper tape, or a floppy; hot start from where it was halted; or warm start from some intermediate state.

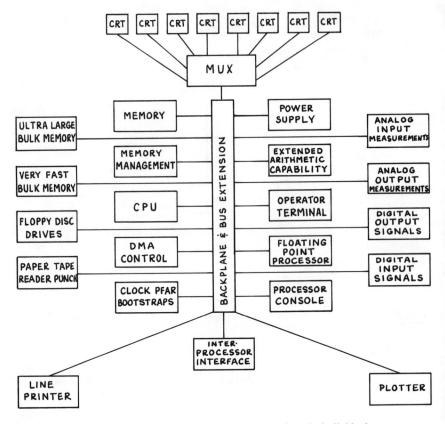

Figure 4-20f. The progress of building a hardware system from its individual components: a fully expanded process control configuration.

To provide all the slots and connections necessary for the final configuration, it may be necessary to provide either a bus extension or an ancillary backplane in an expander box. At this point, the system has a minimum configuration and could be used as an inconvenient, but capable, computing device (see Figures 4-20c and 4-22).

The inconvenience is twofold. There is no way to enter programs other than through the processor console or the operator terminal, and there is no way of storing programs not in use. The addition of a paper-tape reader punch adds a fast way of entering programs and data plus a semifast way of getting them out. A very fast bulk memory which operates at memory speeds with no latency for electromechanical parts adds a convenient way of storing

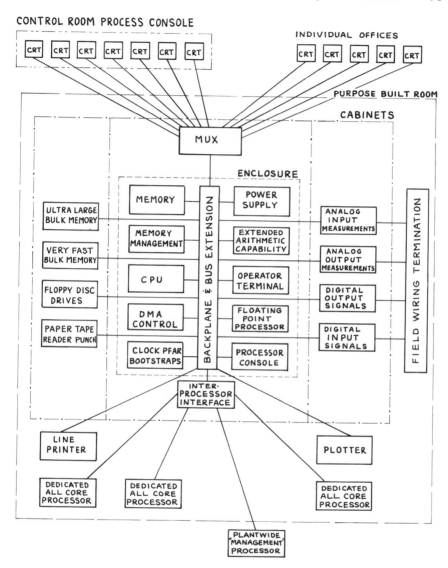

CONTROL ROOM PROCESS CONSOLE

INDIVIDUAL OFFICES

PURPOSE BUILT ROOM

CABINETS

MUX

ENCLOSURE

Figure 4-20g. The progress of building a hardware system from its individual components: a fully expanded process control configuration and its environment.

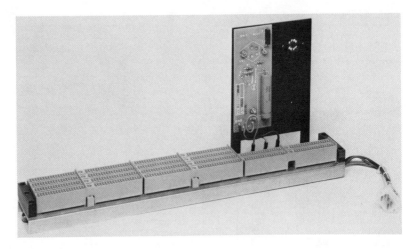

Figure 4-21. The power controller and backplane of a small computer. (Courtesy Plessey Peripheral Systems.)

Figure 4-22. A series of cards (CPU, memory, I/O) in a card cage which, when mounted in a chassis together with power transformers, cooling fans, and an operator's front panel, becomes a viable computer. (Courtesy Digital Equipment Corporation.)

Figure 4-23. The fronts of minicomputers showing the lights and switches that comprise a programmer's console. (Courtesy Data General Corporation.)

programs and data. It is necessary to add a direct-memory access (DMA) controller to the system to coordinate the CPU, the memory, and the very fast bulk memory. The system has now been expanded to a convenient computing device. Programs can be entered either by paper tape or by the keyboard, stored in the bulk memory, and processed at will. But the system still cannot be used for process control. It is necessary to add a "front-end."

This "front-end" comes in two parts: an analog-to-digital converter (an AI) and a two-way digital controller (DIDO). The AI handles the conversion of the various analog measurement inputs from the real world into digital signals for the computer system. The DIDO handles the digital input signals (from on/off switches) from the real world to the computer, and the digital output signals (flipflops and pulses) from the computer system to the real world. The AI/DIDO combination allows the computer to take in real measurements, be aware of which control valves and controllers it can direct, and change their positions as and when needed (see Figure 4-20d).

At this point, the computer system is still not capable of interfacing in a convenient fashion with people. It is necessary to add operator- and engineer-

Figure 4-24. Any system can be enhanced by the addition of components such as a floating-point processor and extended memory control. (Courtesy Digital Equipment Corporation.)

orientated displays. The addition of a line printer enables the system to print hard-copy (paper) reports, but this must be considered a secondary interface. The printing process is slow, and paper output in bulk is incomprehensible. The primary human interface with the machine must be through the superior soft-copy (video) CRTs. The wisest long-range planning is to provide for as many CRTs as there will be individuals looking at the process. This means planning for as many as 32 CRTs, rather than for as few as two CRTs. How many of these terminal stations should support color or graphics, as well as alphanumeric displays, must be known during the initial planning stages. A real-time, multi-CRT interface system will demand that the stations are multiplexed (MUX) into the CPU's bus (Figure 4-25). The reduction in physical size of the interface hardware and the reduction in length and complexity of the software drivers demand it. In fact, it is unlikely that the interface will be successful if it is not bidirectional DMA as well. The availability of real-time displays to many individual viewers simultaneously is crucial to the progress and extent of the overall project's success and is emphatically not to be dismissed in the original planning as a minor consideration.

Figure 4-25. A configuration can have increased communications throughput by off-loading character interrupt processing to a data communications controller. (Courtesy Data General Corporation.)

For real-time displays to be meaningful, they must be rewritten frequently. Consequently, high-baud rates will be required. The screens will need to be rewritten once a second, every five seconds, or every 10 seconds. To support this, baud rates of 9600, 19200, and 38400 are necessary, almost precluding non-DMA techniques. We now have a system that not only can take charge of a process, but one which several people can use simultaneously and from which hard-copy reports can be obtained for conventional information dissemination, the archives, and scratch doodling (see Figure 4-20e).

This complex arrangement of hardware can be made considerably more sophisticated with further enhancements. The addition of a hard-copy plotter will permit the capture of any video graphic desired. Provision of analog output within the front-end will enable the system to drive any desired analog recording or indicating device (or analog controller) or even provide analog signals to other computer-driven systems. A hardware-extended arithmetic capability will decrease computation times considerably and will extend the instruction set of the machine.(Figure 4-24).

A floating-point processor converts the system into a dual-processor combination and drastically reduces computation times. It will ordinarily allow a choice of arithmetic precision as well as provide the automatic generation of many mathematical functions. It can be number crunching while the host processor attends to its other functions—allowing a true duality. An interprocessor interface can connect the process control system with other systems, be they lateral or hierarchical. It can be a simple serial interface running at peripheral speeds, or it can be parallel interface running at memory speeds. The intelligence of the CRT interface MUX can be increased greatly, even to the point of satisfying the bulk of the keystrike requests without host CPU intervention. In can be made to support a mind-boggling 256 stations.

To increase the power of the system even more, the memory can be expanded beyond the capacity of the original instruction set and processor word size, or it can be made faster. This involves adding memory management in one or both of two forms. In terms of size the memory can be expanded four or eight times beyond the original architecture. But only the original architecture capacity is accessible at any one processor cycle. This allows giant programs, massive amounts of data, and even alternative system modes to be concurrently core-resident yet accessible in just one or two extra processor cycles. In terms of speed, cache-memory can be added in limited amounts. Cache-memory operates at processor speeds, that is, 10-20 times faster than normal memory.

Paper tape is a bulky, inconvenient, and slow method of intersystem information transfer. Removable floppy disks are fast and compact. They are also an infinitely expandable way of storing information but do need to be loaded and unloaded by hand. For some systems, they can even serve as the bulk memory.

The final enhancement that can be added to the system is ultralarge bulk memory capable of storing 20, 40, or 80 megabytes of information. A storage capacity of 20 megabytes will allow on-line storage of almost every conceivable display of the chemical process under scrutiny and have room left to store a program library that would take several years to write. Furthermore, 40 megabytes by itself can store the actual 16-bit values of 320 analog measurements made every 10 seconds in a complete week, or those made every 10 minutes in a whole year—a veritable trove of information on the process. A supersystem with all these enhancements is shown configured in Figure 4-20f. A system fully enhanced but not totally expanded may be less than double the cost of the unenhanced version configured for a like number of channels (process and terminals).

The final decisions to be made are how, where, with whom, and with what the complete system is to be packaged, sited, and interfaced. Outwardly, a configured system is compact and plain looking (Figure 4-26). The CPU, the memory, and vitrually all the enhancements will take one enclosure of 21

Figure 4-26. A minicomputer process control system. In the foreground is the desk console containing the peripherals of the operator interface with two large CRTs, a keyboard, a printer, and an array of analog meters. In the background is the cabinetry containing the very heart of the system, the CPU, the memory, and the controllers for the process interface and the communications channels. (Courtesy Rosemount, Inc.)

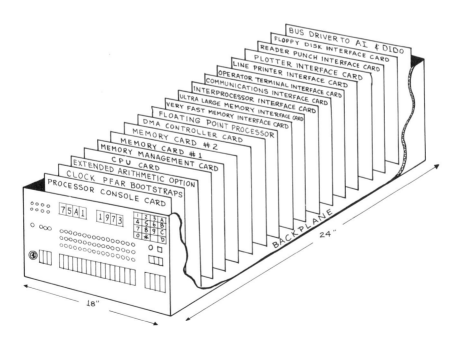

Figure 4-27. An expanded processor in a CPU box with a full complement of enhancements.

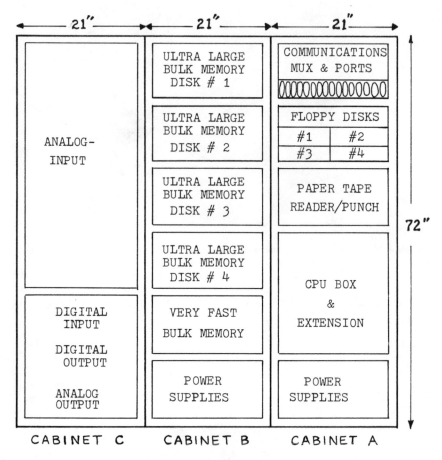

Figure 4-28. An expanded system can be mounted in three cabinets.

inches in a standard 19½-inch wide cabinet (Figure 4-27). The power supply, non-process interfaces, floppy-disk drives, and paper-tape reader punch will fill the remainder of the cabinet. A second cabinet will be required for the bulk memories. A third cabinet will be required for the process front-end, that is, the AI/AO and DI/DO subsystems (Figure 4-28).

A field-wiring termination panel of about 4 x 8 feet will be required for 320 measurement inputs and the supervision/control of 36 control valve loops (Figure 4-29). The three cabinets plus the line printer, the plotter, and the operator terminal will require an office about 14 x 12 feet in size, not necessarily with a raised floor or special air conditioning (Figure 4-30).

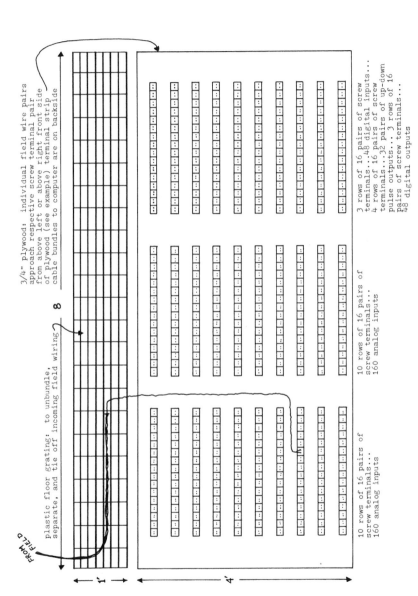

Figure 4-29. A combined field-wiring termination panel and screw terminal assemblies of the computer's front-end.

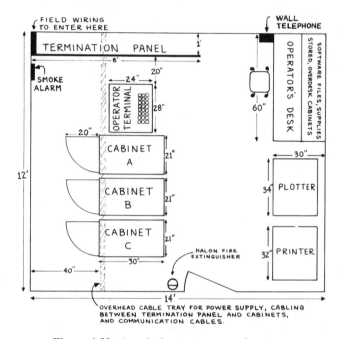

Figure 4-30. A typical computer room layout.

Figure 4-31a. A single-board microcomputer: photograph of actual board. (Courtesy Motorola Inc.)

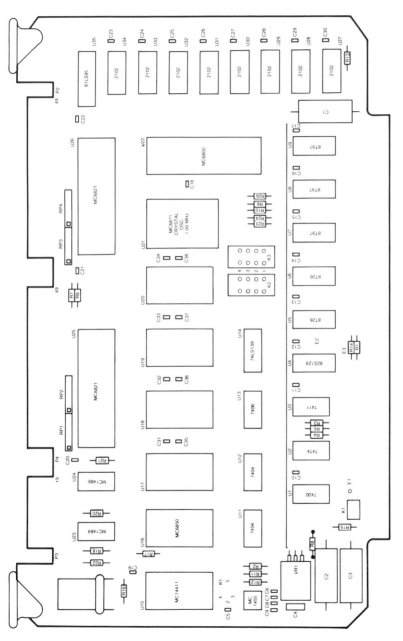

Figure 4-31b. A single-board microcomputer: the parts location diagram. (Courtesy Motorola Inc.)

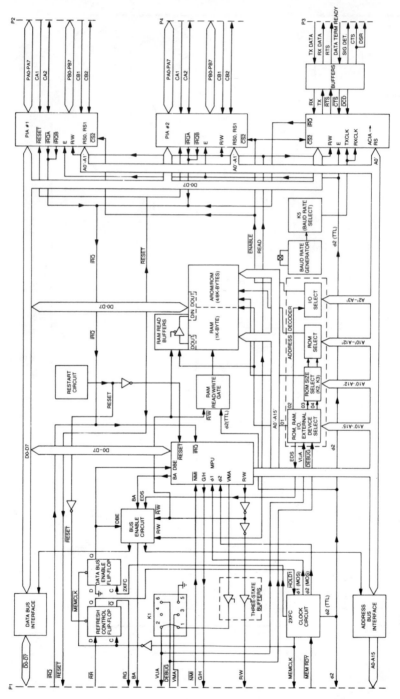

Figure 4-31c. A single-board microcomputer: the block diagram describing functionality. (Courtesy Motorola Inc.)

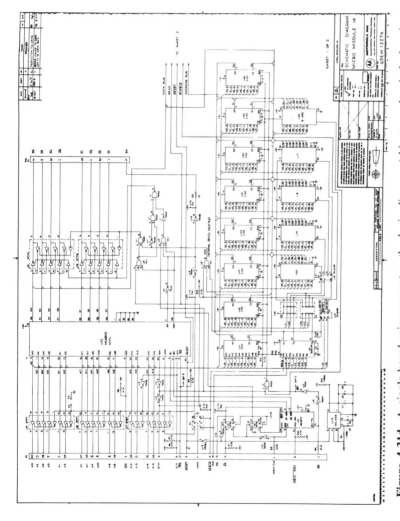

Figure 4-31d. A single-board microcomputer: the logic diagram (akin to a chemical plant in miniature: an aerial view, a plot plan, a process flow sheet, and the P & ID). (Courtesy Motorola Inc.)

For operator interface in the control room, a purpose-built console carrying eight 12-inch CRT screens and two custom keyboards will be no larger than an office desk. Individual office CRTs for engineers, the manager, and the production clerk will take up no more of their desk space than would a small portable television. The interprocessor interface will provide processor-to-processor communication. The process system will direct and monitor the performance of purpose-dedicated, all-core systems doing environment monitoring, process-stream analysis, alarming, and sequencing. The process system will be a reporting unit within a plant-wide system of coordination and information. An overall configuration scheme is shown in Figure 4-20g.

5

COMPUTER PROGRAMMING, OPERATION, AND SOFTWARE

Addresses and Instructions

A computer does not handle numbers directly. A computer deals with numbers much as people deal with liquids. If five gallons are to be put into bucket C, and there are two gallons in bucket A, three gallons in bucket B, and a mixing bucket, the operation will go as follows:

> Empty the mixer. Pour the contents of bucket A into the mixer, then add the contents of bucket B. Put the contents of the mixer into bucket C. Now consider the numbers. To put 98 units into bucket C, there must be 33 units in bucket A and 65 units in bucket B. The mixer is emptied. The contents of buckets A and B are added to the mixer. Then the contents of the mixer are put into bucket C. Result: bucket C contains 98 units.

In a computer the buckets are addresses, and the *adding, emptying,* and *putting* are instructions. If the computer is to be used to add 33 and 65 to make 98, the machine must be loaded with 33, 65, the instructions to "empty," "add," "add," and "empty," and a location provided for the result.

Program

A program is a series of instructions that the computer is to follow. It may contain constants necessary to the calculations as well as the instructions.

Subroutines

A subroutine is a series of computer instructions and operations used either more than once in a program or by more than one program. In short, a

105

subroutine is a miniprogram within a program. It is quicker and easier to detour from the program to the subroutine, and then back to the original program, than it is to nest a repeat set of identical instructions within the program and make the detour unnecessary. If entry to a subroutine is made, then the return address must be stored automatically either in the subroutine or elsewhere in the computer. If that is not done, return to the original program will be impossible.

Datasets

Datasets are sets of numbers. They will not stand alone as programs. They are usually required to support proper programs with data.

Programming Simple Operations

The instructions which guide the computer through any desired computation are stored in the computer as coded digits, as is the numerical data itself. This could allow the computer to modify already-stored instructions, but to do so is a technique used only in advanced programming, and then only as a last resort.

Each word of a program in memory has two parts: the instruction part and the operand part. The instruction part is digital code for whatever action the machine is to do during that instruction cycle. The operand part is an address, or directions to an address, of a number or a register to which that action applies.

A Simple Program

All the arithmetic work will be done in a register called the accumulator. A double letter is the mnemonic address code for the single-letter number, i.e., number A is in address AA. There is a one-digit instruction code and a three-digit operand code. An example instruction set follows:

Mnemonic	Digital Code	Action
CLA	7000	Clears the accumulator, no operand.
ADD	1000	Adds the contents of the operand address to the accumulator.
PUT	3000	Puts the contents of the accumulator in the operand address.
MPY	2000	Multiplies the contents of the accumulator by the contents of the operand address.

continued

continued

Mnemonic	Digital Code	Action
HLT	4000	Halts the processor.
SUB	0000	Subtracts the contents of the operand address from the accumulator.
SZA	5000	Skip one instruction if accumulator is zero.
JMP	6000	Jump to the address in the operand.
JMS	8000	Jump to the subroutine which begins at the address in the operand.

It will be necessary to use one subroutine in the example, subroutine PRINT. This subroutine prints out on the teletypewriter the numbers with addresses stored immediately behind the subroutine call. At the end of the subroutine, a return is made to the third location behind the calling address. PRINT begins at memory location 699 and is known to work. All these concepts are demonstrated in the example program that follows.

Calculate $Y = (A * X + B) * X + C$ for $X = 1,2,3,4,5,6,7$, and print out the values of Y and X for each X, when $A = 5$, $B = 7$, and $C = 41$:

Number	Storage Mnemonic	Number storage required: Initial Value
Y	YY	0
X	XX	0
A	AA	5
B	BB	7
C	CC	41
1	ONE	1
8	EIGHT	8

The program in mnemonic and digital code:

Memory Location	Address Mnemonic	Mnemonic Instruction Operand	Digital Instruction & Operand
1	START,	CLA	7000
2		ADD ONE	1025
3		PUT XX	3028
4	LOOP,	CLA	7000

continued

continued

Memory Location	Address Mnemonic	Mnemonic Instruction Operand	Digital Instruction & Operand
5		ADD AA	1022
6		MPY XX	2028
7		ADD BB	1023
8		MPY XX	2028
9		ADD CC	1024
10		PUT YY	3027
11		JMS PRINT	8699
12		XX	0028
13		YY	0027
14		CLA	7000
15		ADD XX	1028
16		ADD ONE	1025
17		PUT XX	3028
18		SUB EIGHT	0026
19		SZA	5000
20		JMP LOOP	6004
21	FINISH,	HLT	4000
22	AA,	5	0005
23	BB,	7	0007
24	CC,	41	0041
25	ONE	1	0001
26	EIGHT	8	0008
27	YY,	0	0000
28	XX,	0	0000

The resulting output:

1	53
2	75
3	107
4	149
5	201
6	263
7	335

Instruction Set

The computer is organized around a few very simple instructions and their variations:

CLEAR REG R   Clear register R; possibly up to eight registers can be addressed independently.

GET NUMADD, REG R	Get the contents of the address NUM-ADD, and add them to register R.
AND NUMADD, REG R	Logical "and" contents of register R and the contents of the address NUMADD. Leave results in register R.
ISZ REG R	Increment register R and skip the next instruction only if the register has become zero.
DSZ REG R	Decrement register R and skip the next instruction only if the register has become zero.
PUT NUMADD, REG R	Put the contents of register R in the address NUMADD.
JMP NUMADD	Jump to the address NUMADD.
JMS NUMSADD	Jump to the subroutine beginning in address NUMADD, storing the address just left.
SGE REG R	Skip next instruction if register R is equal to or greater than zero.
SEQ REG R	Skip next instruction if register R is equal to zero.
SLE REG R	Skip next instruction if register R is less than or equal to zero.
SNQ REG R	Skip next instruction if register R is not equal to zero.
SGT REG R	Skip next instruction if register R is greater than zero.
SLT REG R	Skip next instruction if register R is less than zero.

Register Bit Manipulation

There is a further series of instructions that allow manipulation and testing of the individual bits in the various registers.

Input/Output Instructions

For each peripheral and bus extension device, there are instructions that pass data back and forth and interrorgate the device on status.

Floating Numbers

In scientific notation 123.45 is written 1.2345*E02. A very similar system has been adopted for storing numbers in a computer. The number is converted into a power of two exponent and a fraction between one-half and less than one. Some examples are:

$$10 = 2^4 * \tfrac{5}{8}$$
$$2 = 2^2 * \tfrac{1}{2}$$
$$1.5 = 2^1 * \tfrac{3}{4}$$
$$1 = 2^1 * \tfrac{1}{2}$$
$$100 = 2^7 * \tfrac{25}{32}$$

Usually, the exponent is stored as one byte and the mantissa as three bytes (see Figure 5-1). This is enough for seven-decimal digit significance.

In practice, only seven bits of the first byte are used for the exponent. The very first bit is used to sign the complete number. An "excess 64" notation is adopted with the exponent. In "excess 64" notation an exponent of zero is carried as a value of 64, an exponent of 63 is carried as 127, and an exponent of –63 is carried as 1. Because the mantissa always will be normalized, a hidden bit convention can be adopted if a way is found to express a zero value for the complete floating number. This is easily done because of the bias applied to the exponent. A number can be defined as zero if the carried exponent is zero and the sign bit is off, i.e., the first byte equals zero.

Not all manufacturers implement identical floating-point notations. Not all conventions normalize individual bits. Some manufacturers normalize half-bytes (nibble normalizing). Because of the differing conventions, some microprocessors carry greater significance in their notation than some mainframes.

A Floating-Point Processor

A floating-point processor is to a computer what a hand-held calculator is to an engineer. Without an extended arithmetic element, a computer does multiplication and division by repeated addition in much the same way as does a mechanical calculator. With an extended arithmetic element, the computer is able to do long multiplication and long division of integer numbers by electronic circuitry in one operation. A floating-point processor carries out arithmetic operations between floating-point numbers independently of the computer and stores the result wherever required. An arithmetic operation by the computer alone between two floating-point numbers may take 1000 microseconds. With an extended arithmetic element, this time can be reduced to 200 microseconds. A floating-point processor reduces the time to 30 microseconds.

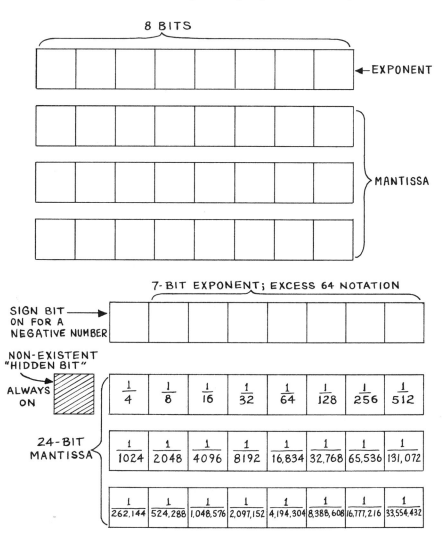

Figure 5-1. Floating-point conventions.

Floating-Point Math Instruction Set

(fp# floating-point number)
(fpa floating-point accumulator)

Arithmetic Operations
    FADD   Add a fp# to the fpa

FSUB   Subtract a fp# from the fpa
FMPY   Multiply the fpa by a fp#
FDIV   Divide the fpa by a fp#
FGET   Clear the fpa and add a fp#
FPUT   Store the fpa where indicated, but do not clear the fpa

Function Generators (Most functions are calculated using optimized Taylor series expansions.)

SQUARE Square of fpa, the result is left in the fpa.
SQROOT Square root of fpa, the result is left in the fpa.
SIN   Sine of fpa, the result is left in the fpa.
COS   Cosine of fpa, the result is left in the fpa.
TAN   Tangent of fpa, the result is left in the fpa.
LOG   Common log of fpa, the result is left in the fpa.
LN   Natural log of fpa, the result is left in the fpa.
EXP   Exponential of fpa, the result is left in the fpa.
FLOAT   Float an integer number into a fp# in fpa.
FIX   Fix an fp# from the fpa to integer form.
NORM   Normalize the fpa.
FINP   To generate an fp# from a stored string of integers that are the individual digits and decimal point of a decimal number in ASCII code.
FOUT   To generate from an fp# a string of ASCII code of the individual digits and decimal point of a decimal number.

Disk Directory

As the name implies, this is a record of all the programs and datasets residing on the disks. It will contain the disk address of where the individual programs and datasets reside on disk, and how much space they occupy. Also, the directory may be expanded to contain the program and dataset names, and any information necessary to accomplish the running of a program in memory.

System Director

To run automatically, a computer must run in a small loop without stopping. This loop does not have to be any more complicated than the equivalent of the human activity of "running in place." Also, the computer must be able to depart that loop, perform any task asked of it, and return to the loop. The tasks that it must accomplish will either be handling the hardware or executing a series of programs. The system director exists to handle the tasking. It is the overall coordinator and arbitrator for:

1. Input/output processing
2. Interrupt handling
3. Disk file management
4. Console-directed operations
5. Keyboard requests
6. The clock and calendar
7. Acknowledgments
8. Data transfers between devices
9. The batching of programs
10. The display of processor conditions and states

Interrupts

At 9600 baud, 960 characters per second are sent from the computer to the peripheral, or one every millisecond. If the sector from the last track on a floppy is required, it may take 450 milliseconds to find it. Even on a hard disk with fixed heads it may take 25 milliseconds to read the desired record. If the computer is capable of executing instructions at the rate of one every two microseconds, then, while waiting to send the next character, it can execute 500 instructions; while waiting for the floppy sector, it can execute 225,000 instructions; while waiting for the record, it can execute 12,500 instructions.

Interrupts enable a processor to run programs as a foreground task while manipulating the peripheral hardware as a background task. For the character transmission, 10 instructions are sufficient to find and send the next character. The computer then processes a further 490 instructions of the program. It is interrupted by the peripheral's interface announcing that it is "clear" and ready to send the next character. The 491st program instruction is put aside, and its place marked, then the computer finds and sends the next character before picking up its place again. When the sector is found, the floppy-driver interface will announce it, the program will be interrupted, the floppy serviced, and the program resumed. When the desired record is read into memory, the hard-disk interface will announce it, the program will be interrupted, the disk serviced, and the program resumed.

Once in every machine instruction cycle, the interrupt bus is tested. If it is asserted (i.e., some interface is trying to announce something), then the processor begins an interrupt search routine. The current "program counter" and the values of the various registers are stored. The processor then interrogates each device until it finds the interface that asserted the interrupt bus. The processor then does whatever is necessary. If no more interrupts are pending, the program counter is restored, the original values of the registers are restored, and the computer returns to processing the program.

In the system configured in the previous chapter each of the following devices was capable of making an interrupt:

1. The clock
2. The power-fail detection device
3. The floating-point processor
4. The very fast bulk memory interface
5. The ultralarge bulk memory interface
6. The interprocessor interface
7. The communications interface
8. The operator's terminal interface
9. The line printer interface
10. The plotter interface
11. The paper-tape reader punch interface
12. The floppy-disk interface
13. The AI
14. The DIDO

System Software

How comprehensive does the supplied software need to be? It has to be comprehensive enough to allow the owner/user to turn the power on, hit the start button, enter the name of the file he is going to build, and enter whatever he wishes to be the contents of the file. The supplied software must create a permanent storage place automatically for that file. It must permit the file to be listed back if the owner/user so wishes; it must execute the file's contents on request. Finally, the supplied software must be comprehensive enough to suffer the power being turned off without ill effect to the machine's future performance. Do not buy a machine that comes with software that is any less powerful.

Controlling a chemical process with programmable processors involves using programmable processors for two distinct purposes. Therefore, if two machines are used, there are considerable benefits to the chemical process and chemical plant: convenience, security, safety, and machine accessibility.

The first machine is used to prepare and test process control programs. The second machine is used to control the process plant itself with the prepared and tested programs. One machine can be used for the two purposes. It is arguable whether one machine or two machines is the more cost-effective approach if only direct out-of-pocket expenses are considered. If all the opportunity costs are recognized, then the two-processor approach is definitely more effective. For this reason, the Witches' Brew Process will use two machines. The first, "Tweedledee," will be used for program preparation, program testing, and off-line number crunching. The second machine, "Tweedledum," will be the process-dedicated process controller.

Each of the two machines has two sets of complementing software: one set of software coordinates the machine's function according to the machine's purpose; the other set of software generates the benefits to the owner of the machine's purpose. Whichever machine is being considered, whichever set of software is under study, the software exists in discrete, uniquely named files such that one file contains either a set of data or a program. For Tweedledee, the coordinating software has been supplied by the manufacturer and is specifically written to allow creation and modification of files. The benefit-generating software will be the custom-written programs for Tweedledum. In the case of Tweedledum the coordinating software is specifically written to coordinate the machine and its peripherals for real-time process control (a very demanding purpose), and the benefit-generating software will be the process control algorithms unique to the Witches' Brew Process.

In system software the custom has been to develop an operating system purpose-written for a minimal configuration of hardware, a very limited number of users, or a few undemanding simultaneous purposes. Demanding users who have expanded their memory to maximum size, added a floating-point processor, extended their disk storage to gargantuan proportions, and have a full complement of peripherals; or who wish to timeshare with 20-30 4800-baud-plus terminals; or who must have videographics, solve various linear programming problems, and do real-time statistical analysis simultaneously are catered to by extensions, patches, and alterations to the standard operating system. The demanding situations get a stretched version of a noncomprehensive operating system.

What really is demanded, of course, is that the operating system be purpose-written for the full configuration of hardware (enhanced hardware, of course) with a deliberately large number of users in mind, who in turn would wish to run their most complex programs all day long in real time. Users who do not fully configure or enhance their hardware, or timeshare only amongst a few and then only with simple programs, and are not inconvenienced by long turnaround times on batch jobs receive a detuned operating system, secure in the knowledge that future expansion works. This cannot be guaranteed of today's restricted-vision operating systems.

Languages

Use BASIC for developing and testing ideas on form of outputs, moderate number crunching, once-only projects, and interactive programs. Use FORTRAN for the repetitive projects that involve heavy number crunching and that can be batched. Use Pascal (or other structured languages) for system development and the development of user programs of a real-time nature. Use assembly language wherever necessary to optimize the operation of BASIC, FORTRAN, or Pascal programs. Do not consider any one language a universal panacea for all the problems you will encounter in process automation by programmable processors. You will have enough difficulties; you do not have to create them.

Software: Tweedledee—The Development System

System Director

A system director optimized for interactive communication over a few channels, batch processing, and unbounded universal file handling. It need not be real time nor a multilevel system. The system will have a minimal front-end.

Utility Programs

These are delivered with the system director and are guaranteed to work as follows:

CAT.   A directory of the files ("run" documents, "source" documents, "methodology" documents, and datasets) within the system.

ERASE.   The removal and expunging of a file.

COPY.   The duplication of a file.

RENAME.   The renaming of a file.

BUILD.   The creation of a new file.

EDIT.   The modification of an existent file.

RUN.   The execution of a program from "run" or "source" documents in such a fashion that errors within the program are trapped and terminate the program's execution without affecting the system.

DEBUGGER.   The execution of a program from "run" or "source" documents in such a fashion that single-instruction execution at various breakpoints and corrections can be made on an interactive basis through a terminal without terminating the program's execution.

TRANSLATION.   The creation of a machine-coded relocatable object file (the "run" document) from ASCII-coded source files (the "source" documents).

DISASSEMBLE. The reverse of TRANSLATION.

SWAP. The moving of files back and forth between different mass-memory peripherals, changing the file format as and when necessary.

HELP. A memory-tickling output display tailored to whatever mode the computer is in.

LIST

LOAD. The listing/loading of either "run" documents or "source" documents through entry/exit-type peripherals: floppies, paper tape, magnetic tape, CRTs, and line printers.

CONFIG. The capability to reconfigure any of the system's attributes on line.

PACK. The capability to shuffle all the files on a mass-memory peripheral into adjacent storage locations, as well as the capability to nest several short files within one holding file.

User Programs. All the programs on Tweedledum (their "methodology," "source," and "run" documents), various data-interpreting programs, the macroprocess control programs, the program that facilitates the creation of the display dataset-cum-tables for Tweedledum, and any other programs the owner may find beneficial to have and to use.

Software: Tweedledum—The Process Control System

System Director

A system director highly optimized for a real-time environment and the selected configuration. It will be multilevel. The system will have a huge front-end and a large communications interface. There will be bounded specialized file handling.

Utility Programs

LOAD. The loading of "run" documents only through entry-type peripherals. Load will erase and pack as necessary.

PATCH. The modification or patching of "run" documents on-line.

DISPLAY. The creation of a display on the CRT on request.

REFRESH. The updating of all current displays.

REPORT. The printing of reports.

REQUEST. The handling of any keyboard request.

FRONT-END. The periodic measurement and conversion of all incoming analog values to engineering values in floating-point notation; the transmission of pulse-train adjustments to the regulators; the monitoring and update of all the statuses of the controllers and regulators.

User Programs.   Whatever programs and displays the owner considers necessary to monitor and microcontrol his process. The system will be loaded with dummy programs and dummy displays as necessary to maintain the dedicated file-structure integrity. Chapter 6 describes Tweedledum's user software.

Program Development

Making a computer produce worthwhile results is a grand achievement because it is such a complex task. Obviously, to produce from the machine an ordered result, the machine must have followed an equally ordered list of instructions—a list that contained neither errors nor flaws. The list is the embodiment of some person's perception of how that machine must take in the data, massage it, and put out the results. If either the results or the form of the results is not what was wanted, then the person's perception was wrong, and the list must be altered. There is a trial-and-adjustment seesaw between the perception of how to make the machine produce the results, (as that perception exists in the list), and the actual results or form of the results themselves. If the results do not match the original desires, then the perception is altered, the list is manipulated, and another attempt is made to produce the required results. We can make this into a formal progress. A result is desired. An acknowledgment is made that the machine can be used to effect the desired result. A methodology is formulated. This methodology describes how to use the machine for it to produce the result. That methodology is transcribed into a disciplined syntax. The syntax is converted to machine instructions. The machine, in a trial, is so directed into producing a result. Then an analysis must be made. Are the results satisfactory? If "yes," the progress is complete. If the results are not satisfactory, could syntax changes produce the desired result? If syntax changes could not produce the correct results, could a change in the methodology? If this could not produce the desired results, could there be a change in the concept of what constitutes a desirable result? Then whatever changes are needed are incorporated into the perception and further trials are attempted.

Repeated circling of the progress, from perception to trial to analysis, will eventually result in an ordered list of instructions that engenders the desired results. Such a progress is programming in the fullest sense, and the perception is what is called a program.

A program has three parts: a methodology document, a source document, and a run document. In bold terms the methodology document describes the program in non-specialized terms and can be read and understood by the intelligent lay person. It is either a logic flowsheet or a written description. The source document is an unabridged version of the program written in pseudo-computer instructions and pseudo-algebra, liberally sprinkled with comments in plain English. The document can be read and understood by people who have learned the particular syntax used. The run document is a compact list of the machine instructions that constitute the program. It usually exists only in machine-readable form, and is devoid of all but the essentials. Only the most highly skilled can decipher this document.

If we take a closer look at the analysis, we can establish eight kinds of intermediate causes as to why a trial of a process control program might not produce the wanted results:

1. Typing errors in the source document

2. Incorrect program syntax

3. Program logic errors

4. System software restrictions

5. System hardware restrictions

6. Ohdamns

7. Process instrumentation problems

8. Chemical process and plant problems

Generally, failure attributable to the first three causes (typos, syntax, logic) will be uncovered in attempts to generate the run document. A working run document will generally be needed before attempts can be made on the next three causes (system software, system hardware, ohdamns). A working system and a compatible working run document are needed before you can discover what brickbats the real world has hidden. Instrument problems will have to be solved even before you get to the problems of the process. When the methodology document is accurate, the source document correct, the run document perfect, the system working, the ohdamns long forgotten, the instruments exact, and the plant right, you can still get results if—and only if—people want those results badly enough to close the necessary switches and give you user use.

If at any one of the 10 stages between conception and user use there is an error within the program or a failure of the program to work as envisioned, the redevelopment of the program will begin at stage one (Figure 5-2). After the program has been modified, each stage must be passed again, whether or not it has been passed before. Program development is tedious, the more so because complete attention to detail is demanded. There are no shortcuts; each and every stage must be passed (Figure 5-3). Experience will make the overall process go faster. The process also will go faster if a program is broken down into self-contained subsections. It is quicker to pass the several parts of the whole through the development process than it is to attempt to pass through the undivided whole.

STAGE & ACTIVITY	MILESTONES	PROBLEMS
	CONCEPTION	
STAGE I CONCEPTUALISING drafting conceptional program		
	METHODOLOGY DOCUMENT	
STAGE 11 EDITING typing program into the machine		
	SOURCE DOCUMENT	simple typing errors, misspelling
STAGE III COMPILING correcting compilation errors		incorrect data, addresses, psuedo ops, syntax
	SUCCESSFUL COMPILATION	
STAGE IV TRIAL RUNS:DOCUMENT ERRORS correcting run errors		missing code, wrong code program logic errors
	PRELIMINARY RUN DOCUMENT	
STAGE V TRIAL RUNS:SYSTEM SOFTWARE RESTRICTIONS overcoming system software restrictions		non-existent library routines, historical data not available, program too large, program too slow, I/O buffers poorly sized
STAGE VI TRIAL RUNS:SYSTEM HARDWARE RESTRICTIONS overcoming system hardware restrictions		serial interface not available, poor communications protocol, disk access too slow, pulsing too fast
	WORKING RUN DOCUMENT	
STAGE VII INITIAL SOFTWARE TEST correcting ohdamns		transposed logic, wrong variables wrong formulae, incorrect factors

STAGE VIII INSTRUMENTATION TRIALS   PRELIMINARY REAL RESULTS
    correcting instrument problems
                                           cannot make anticipated measurements,
                                           analysis times too long, inaccuracy,
                                           non-repeatability, unreliability,
                                           sticking valves, wild measurement
                                           fluctuations

                    COMPUTER CONTROL POSSIBLE

STAGE IX CONTROL LOOP TESTS
    correcting process & plant problems
                                           residence times, imperfect mixing,
                                           insufficient pump horsepower,
                                           incorrect piping, restricted heat
                                           transfer

                    VIABLE COMPUTER PROCESS CONTROL

STAGE X TRAINING
    training people to use system
                                           fear of the unknown, unwillingness
                                           to adopt new work practices,
                                           unwillingness to learn new technology

            USER USE

        FAILURE IN APPLICATION   ANTICIPATED RESULTS

Figure 5-2. The progress of a "process-control" program.

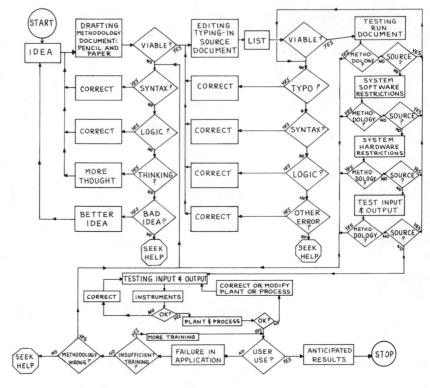

Figure 5-3. The flow chart of the 10-stage process of developing a program.

The concept of a 10-stage developmental process is true for any program whether it is a management report, a control algorithm, a video display, or an operating system. It is also true whether it is a brand-new application, a minor modification to a working program, or a revamping of old software. In the following examination of the 10 stages a control program is being developed.

Stage 1. Conceptualizing

What precisely is the program to do? Let us define the problem, being very clear about the purpose of the output from the computer. Write down clearly the whole concept of the program. This serves two purposes: the written concept becomes a methodology document for later reference by yourself or others, and it provides the skeleton upon which you will hang the various subsections of the ensuing program. This is the time to put in a lot of thought. Erasures, alterations, expansion, reordering—these take very little effort and almost no time, while the concept is being written in pencil on paper. Any time gained by rushing through the first step surely will be lost in delays at later stages. Take your time during Stage 1. Be sure, during this conceptualizing, that all possible eventualities can be handled by the program. Expect people-errors and equipment malfunctions, and create programs that can handle them. Go over the draft proposal step by step. If the program is to make decisions on massaged data, go through the proposal with both higher-than-to-be-expected and lower-than-to-be-expected values. Only when the proposed program works on paper and looks as secure as you can make it are you ready to take the program to Stage 2.

Stage 2. Editing

The rough draft is now typed into the computer. It is a mechanical process and almost painless with good time-sharing software. Any typos that are recognized as such at the time of entry are erased easily, usually by one backspace for each incorrect keystrike. For typos uncovered in earlier work, either the whole of the offending line is reentered, or the incorrect section is selectively edited. Inevitably, the typing-in of the program will be accompanied by undetected typos, misspelling, incomplete deletions, forgotten additions, misplaced insertions, and missing characters. When the last portion of the rough draft has been entered, have the whole program typed out. Correct any errors you observe. Listing the program at this stage generally is not done. Usually time-share terminals run at slow print rates and people prefer to compile immediately. It costs both computer resource time and your own time to use a compiler to find your editing errors. Either list through a high baud-rate terminal or list through the system's high-speed line printer. Eliminate all visible errors, then compile. Avoid premature compilation. The source document is now ready for Stage 3.

Stage 3. Compiling

This is the conversion of your source document into machine instruction code—the run document. Anything that the machine does not consider legitimate for the language you are using in your source document is flagged as an error. Such errors can be undetected typos, entry errors, incorrect language syntax, missing program steps, incomplete data, non-existent addresses, and even misspelled pseudo-ops. All of the errors detected will have to be corrected before the compiling process can conclude successfully. Obviously, editing the source document can and often does lead to new Stage 3 errors. When the program can be compiled without errors, you are ready to proceed to Stage 4.

Stage 4. Trial Runs: Document Errors

It is possible and certainly very likely that the program (even when it is free of compilation errors) will not run to its proper conclusion. The system may abort it, it may go off into the weeds, or it may cause the computer to halt. However, if the program generates output, then both its form and quantity give a useful insight as to where in the program lies the trouble. By shortening the program, carefully including waits or temporary halts, and providing temporary output of a form like "reached statement 22," the problem code can be unearthed. The concentrated attention brought to bear on each part of the program in turn will reveal missing code and wrong code, both of which can be frustratingly syntactically correct. Worse still, you find that under very close examination, some of your best thinking is logically incorrect. When all the problems have been found, corrected, and Stages 2 and 3 successfully renegotiated, the program should no longer stop the machine, go wild, or abort. You now have a run document that, in its preliminary form, works.

Stage 5. Trial Runs: System Software Restrictions

Once the program is compiled or assembled correctly, it still may not be possible either to run it or get worthwhile results because of system software restrictions. Perhaps you have called for a non-existent library routine. Perhaps the historical data you had hoped to manipulate with the program is much greater than the file-handling capabilities of the system software. Perhaps your program is too long for the available space and cannot be overlaid. Maybe the system input/output buffers are far too short to hold the requisite number of characters. It could happen that the system software will neither accept nor issue some of the character codes you wish to use. There are only two choices in solving system software restrictions: either change your program to suit the system software available, or get a systems programmer to alter the system to suit your program. Changing your program does put you back at the beginning of the development process, but it is less complicated than having the system altered. System software alterations are done efficiently and effectively only by the most experienced and knowledgeable of programmers, who, if they oblige you, open up a can of worms. Their altering the system may well give the greatest progress overall, but it is fraught with difficulties. They will have with their system software the same problems you face with your program. The better alternative is to modify your program, unless you are very persuasive or have made system software modifications before.

Stage 6. Trial Runs: System Hardware Restrictions

The original system software was written and designed with the hardware available at that time. Both software and hardware are being developed and improved continuously. But, quite obviously, it will always be that the software development is restricted by the limitations of the currently affordable hardware, rather than the other way around. So existent software usually lags behind available hardware. Your program is an example. To have reached this point, you must either have modified your earliest visions of the how and the what of the program (vis-à-vis the hardware you wish to use), or you have cajoled somebody into modifying the system software so that you can exploit the features of the new hardware of your choice. Programs rarely get through to this stage without fresh thinking.

Now you are about to tackle the hardware restrictions. Implicit in your program are assumptions about the performance of the hardware, that is, you imagine it will perform exactly as you think it will. Beware! The printer may not print out the answers as fast as the measurements are taken. Perhaps the analog controllers are set at the factory to respond to a faster pulsing rate than your machine can generate. Perhaps your disk accesses, slow by nature, are so frequent that the performance of the whole system is degraded, and your program cannot work as intended. In another instance, it could be impossible for the analog measurements to be resolved at the accuracy your calculations require. Or it could be as simple as the peripheral requiring a parallel interface, and your processor only has serial ports. Whatever it may be, you have two choices. Tailor your program to the limits of your current hardware, or modify or replace your hardware to suit your program. People with limited knowledge of their system hardware are wise to change their programs, not the hardware.

Stage 7. Initial Software Test

The run document is viable. It can be run on the system without problem. Now is the time to process various material by the program to check that the program can handle input and properly process it through to output. First, it must be successful with dummy output. Second, the program must handle test input correctly. With correct output for test input, program runs with real input can be tried. Finally, when correct and real output for real input can be generated, the program can be considered to be fully debugged. During the previous process, it is frequently found that the methodology document and source document contain real bloopers. Usually, the harder it is to find a bug, the sillier the blooper is found to be. The logic can be transposed, wrong variables can be included in the program, the wrong formula could have been used, or a factor is found to be incorrect. You have a program that calculates $3X - 2Y$, when in fact you wanted $2X + 3Y$. Or an iteration that you thought would take four or five passes in fact takes 627. Perhaps the report you are formatting looks unexpectedly crowded in the top left-hand corner and will have to be formatted again. You go through a DO loop 30 times instead of 31 times. There can be (and frequently are) a thousand mistakes in a run document that runs on the system without a problem. The mistakes are not typos, language syntax errors, or program logic errors. But what you have successfully programmed the machine to do is not what you intended the machine to do. All the modifications to the program necessary to get your intentions implemented must be made. These modifications are susceptible to any of the errors for which the program has already been tested; therefore, the program will have to go through these stages once again. The first seven stages of program development can be accomplished on the second machine, without interruption to the process machine. At the conclusion of Stage 7, the program "runs as envisioned."

Stage 8. Instrumentation Trials

You now have a program that is able to take input data, massage it, and give output data in a form that is perfectly acceptable to you. The program runs without bombing and within the limits of your system hardware and software. Up to this moment, the program has not worked with real process inputs, and all the effort has taken place on the non-process machine. Now, the run document is loaded into the process machine, and it is run with real-world instrument inputs. It is possible that several problems will surface at this point. The system cannot make the anticipated measurements. Stream analysis may take far too long. Some measurements may be hopelessly inaccurate; others may show an unexpected non-repeatability. The process sensors may be unreliable, have drift problems, or become such a maintenance problem that they are abandoned. The control valves may stick or be so oversized that even the smallest movement sends the process into wild fluctuations. It may be several days or even weeks until all the instrumentation problems are solved in a satisfactory manner.

Never ever modify the software to get around instrument problems that are correctable. Always correct the instruments even if this means reorientating people's thinking. With instrumentation, it is easy to slip into an inextricable mess of kluges, patches, and temporary expedients that will bury the project if anything less than 100% right is accepted. "Almost right" is unacceptable. This is an attitude that has to be assumed by all the people associated with the process if their computer project is to succeed.

Stage 9. Control Loop Tests

In this stage the chemical engineering assumptions inherent in the program are put to the test. Perhaps the wrong formula has been used; incorrect factors may have been entered; valve operation logic may be transposed (opening it instead of closing it); the flow may be streamline, not turbulent; the heat-transfer coefficient may have been overestimated. Only by double- and triple-checking, questioning, and investigating everything that is unusual and un-expected can the subtle errors in the program be found. In fact, it is possible to find errors in programs even after they have been in use for many months, because unusual circumstances not previously encountered produce unex-pected results. Fortunately, errors at this stage are not catastrophic to the program because it is unlikely that correcting them will cause system software or system hardware problems. But it may take more than one attempt to return to this stage with the corrected program.

At this point in the development of the original idea we have real-world measurements as input to the computer, which in turn puts out changes to the real world. The questions that have to be answered before we can progress beyond this stage are: (1) Do these changes to the real world produce the expected results? (2) Have we moved closer to the optimum? (3) Have we reduced our penalty errors?

If the answers are "yes," then we can move to Stage 10. If not, then can we make physical changes to the plant, or can process modifications be made? If this is impossible, then the methodology document must be revised and the *whole* process repeated.

Stage 10. Training

The hardware is reliable, the software is debugged, and the instrumentation is calibrated. Under carefully controlled circumstances, the anticipated re-sults are forthcoming. The chemical plant itself responds to the incremental changes made.

The question now becomes: "Can the computer work its magic if some of the carefully controlled circumstances are relaxed?" If the program environ-ment is changed from "friendly and considerate" to "hostile and abusive," will it still produce the anticipated results? The emphasis moves from the documentation and the program to people—to the people who will use the program in their daily work. They must be trained in the program's use. This may even mean giving them a basic education in computers and programs.

There is an elemental statement of fact (but one that is often overlooked): the program will only produce results if it is used. It will be used if, and only if, the people who use it WILL benefit from its use. They must understand and appreciate the benefits. They must receive education and training in how to use the program. It is better to have the program used voluntarily (or better yet, its use demanded by the beneficiaries) than used under compulsion. Remember—it is a *team* project. If the program is used, then (and only then) can it be called successful. Its development is complete. User use produces the anticipated results.

If the very people whose jobs can be made easier by the program avoid its use, then development is not complete, and there is failure in application. The program must be changed to make its use more attractive or to make the results more compelling. The unwillingness to adopt new work practices or learn new technology must be overcome by further training and education. The fear of the unknown must be eliminated.

It might be that under universal use the program does not generate the anticipated results. The carefully controlled circumstances under which the program was first released are the only conditions in which the program is successful. In this case the methodology document must be revised and certain provisos added to the program's use until such time that a more powerful program can be released that overcomes the objections of the users.

This concludes the examination of the 10 stages. In summary:

IDEA
An error-free methodology
An error-free source document
An error-free compilation
An error-free run module
Correct real-world input to the machine
Correct output from the machine
The ability to affect the real world with that output
Getting benefits from those effects
Getting others to generate those benefits
SUCCESS

Anything less at any stage results in failure. The requirements are knowledge, time, patience, and energy.

Detecting Faulty Hardware

No mention has been made in the foregoing of hardware errors. On any minicomputer or microprocessor, the worst possible thing that can happen during program development is a hardware fault. This will be diagnosed initially as a software problem, putting the software under close scrutiny. The most recent program modifications will be reversed, with the puzzling result that earlier versions of the program which once ran successfully no longer do so. Use this as an indication that something is not right with the hardware. Run diagnostics, or turn the exercise into one of hunting the hardware bug. Once the fault has been found, have it corrected, and return to developing the software. Unrecognized hardware failures during software development can be a serious setback in any project's timetable. The easiest way to prove a suspected failure is to rerun a "known" program. If it fails, the hardware failure is confirmed. If the program does not fail, then the hardware failure is not confirmed, and a more rigorous test must be made with programs that are known to stress the suspected area of failure. Do not ever back off on program modifications because they cause unexpected problems. Rule out hardware errors first. And if you wish to remain popular with field service or the maintenance folks, do the ruling out yourself. They hate to have to write up a service call, and bill you for your own software boo-boo.

Group Effort and Individual Responsibility

It takes many people working together to get a chemical process automated by a programmable processor. Each person has a responsibility to the others not to waste the efforts of the group. This means that the electricians must do an excellent job of wiring and grounding; the instrument technicians must calibrate carefully; the engineers must apply the right theory correctly; the programming must be efficient and error-free; the operators must use the installation—if they do not, they must be explicit as to when, where, and why.

Judging an Installation

There are five ways to judge a chemical process computer installation's use:

1. How many of the loops are closed on the computer early on a Monday morning before supervision arrives—all, nearly all, some or none? How does that compare with the number closed on a Friday afternoon? How many closed loops does the computer in fact support?

2. If on a Sunday at 0100 hours a loop cannot be closed, or if an input measurement fails, or if a peripheral dies, is anybody called out? Does anybody care?

3. How widely disseminated is the understanding of the programs and the system structure? Is it concentrated in one person? If he resigns or goes on an extended leave, does it convulse the system?

4. Is the present system configuration less than it has been in the past or has it never been as powerful? Is it growing or contracting? (A long-term judgment)

5. Does even the most humble operator suggest to the system programmer a program or display that would make the operator's job more rewarding? Are users demanding changes and requesting a greater involvement of the programmable processor in their jobs?

Prescient managers can recognize from the gist of the questions what the acceptable answers should be.

6

COMPUTERIZING
THE WITCHES' BREW PROCESS

When a process is to be computerized, first spend some time thinking and collecting information. As the process supervisor, you must think about what information you need for effective process management, including which routine decisions could be automated, which conventional process control loops give problems, and which inspection records, overtime records, operating procedures, and other items could be put on the computer. As you think over these items, make notes. Write down all your requirements.

Collect examples of each of the log sheets presently used to record the operation of the plant. Read through the operation's logbook, the shift foreman's log, the past maintenance schedules. Make notes of all items that frequently appear. Obtain and reread a copy of the plant's operating manual. As you read through it, annotate your copy in red pencil wherever the manual and current operating practice differ.

As an example of first thoughts, consider the following from the Witches' Brew Process:

1. Tank inventory
2. Automatic gaging of all the tanks every two hours
3. Calculation of inventories
4. Daily report giving opening and closing inventories and inventory change
5. Tank transfer comparisons (loss versus gain)
6. Monitor all tanks for high levels
7. Indicate when it is time to reorder raw materials

8. Daily raw material usage, utility usage, production

9. Monthly raw material, utilities, and production

10. Instantaneous mass balance

11. Instantaneous energy balance

12. Midnight reports

13. Yields, unit ratios, energy ratios for the shift, for the day, and for the month

14. Calculate the variable costs for the shift, for the day or for the month

15. Maintain reactant stoichiometric ratios for changing raw material conditions

16. Maintain reactor temperature and proper heat soak

17. Maintain precise pH control on neutralizer

18. Reduce solvent-to-solute ratio on the extractors

19. Prevent overheating the evaporator

20. Maintain overhead and bottom qualities on the distillation column for the least reflux and the least reboiler steam consumption

21. Printout hourly log sheets with a precise record of all the important variables

22. Maintain records on liquid chromatograph analyses

23. Maintain overtime records

24. Yesterday's, today's, and tomorrow's maintenance schedules on the computer

25. Inspections schedule with memory tickler

26. Certain alarm conditions

27. Checklists

Second, generate a list of how many different displays are required, what datasets are required to support them, and the different programs required (see Tables 6-1, 6-2, and 6-3). Level 1 programs are high-priority programs (to be run immediately after they are queued). Level 2 programs are middle-priority programs (to be run within a few seconds of being queued). Level 3 programs are low-priority programs (to be run as and when possible, usually within minutes of being queued).

Third, size the computer system required. How many inputs to the computer are required to achieve all the items previously listed? (See Table 6-4.) How many variables must be calculated? How many CRT stations will be required? (See Table 6-5.) How many control loops do you propose to close

Table 6-1
CRT Displays for the Witches' Brew Process

A—INDEX
B—ANALOG INPUT CHECKING
C—TRIPLE PRINT AND PLOT
D—REAL-TIME PLOTTING
E—REAL-TIME TRACKING
F—BATCH PLOTTING
G—BATCH TRACKING
H—I-V INVESTIGATION
 I—M-V INVESTIGATION
 J—CONTROL SUMMARY
K—KEYBOARD INSTRUCTIONS
L—PERFORMANCE
M—ALARMS
N—TANK INVENTORY
O—Y'DAY TANK INVENTORY
P—WITCHES' BREW ECONOMICS
Q—DIRECTORY-PROGRAMS
R—DIRECTORY-D&D
S—ENERGY BALANCE
T—MASS BALANCE
U—PROCESS CONDITIONS—REACTORS
V—PROCESS CONDITONS—FINISHING
W—TODAY'S DAILY DATA
X—Y'DAY'S DAILY DATA
Y—7-DAY PERFORMANCE
Z—YTD PERFORMANCE
@—GRAPHIC TRENDING

Note: extra displays could be assigned lowercase notation.

on the computer? (See Table 6-6.) From this information, our Witches' Brew superintendent has decided to purchase a system capable of:

1.   320 analog input measurements

2.   Digital output to/from 32 control valves/controllers

3.   16-channel terminal interface

Projected Costs/Projected Savings

With the system sized and an idea formed of the hardware and software which will be required, the economics can be reviewed. Obviously, no in-

Table 6-2
Datasets for Tweedledum Software Package

#1. COMMON: 1024 VARIABLES
#2. NAMES: DESCRIPTION OF EACH VARIABLE
#3. REAL-TIME PLOTTING DISPLAY ALTERNATIVES
#4. REAL-TIME TRACKING DISPLAY ALTERNATIVES
#5. BATCH PLOTTING DISPLAY ALTERNATIVES
#6. BATCH TRACKING DISPLAY ALTERNATIVES
#7. I-V INVESTIGATION DATA
#8. M-V INVESTIGATION DATA
#9. BATCH PLOTTING DATA
#10. BATCH TRACKING DATA
#11. DVAL/EVAL GAIN & SUBROUTINES
#12. ALARM SET POINTS & MESSAGES
#13. PROGRAM QUEUING DATA
#14. TRENDING DISPLAY ALTERNATIVES
#15. TRENDING DISPLAY DATA
#16. PERFORMANCE DATA HISTORY
#17. TANK DATA
#18.
#19.
#20. LAST 24 HOURS CYCLE-BY-CYCLE HISTORY
#21. CURRENT DAILY DATA AVERAGES
#22.
#23.
#24.
#25.
#26.
#27. LAST MONTH'S DAILY DATA
#28. LAST YEAR'S MONTHLY DATA
#29. THIS YEAR'S MONTHLY DATA
#30. THIS MONTH'S DAILY DATA
#31. EQUIPMENT INVENTORY
#32. INSTRUMENT INVENTORY

---

vestment decision can be made until the costs and savings of the proposal have been calculated, mulled over, and inspected. The projected costs fall into four categories: (1) the process-dedicated computer equipment, (2) the non-process-dedicated computer equipment, (3) the software and training, and (4) making the instrumentation computer-compatible. The projected savings fall broadly into two classes: passive and active. The first class of savings are those savings that accrue before the first computerized control loop is closed. Savings of the second class are those that are directly attributable to process improvements created with the computer actively supervising and controlling the process. With the savings of both classes and the related investment costs available, the entire Witches' Brew Process economics can be reexamined.

Table 6-3
Programs for Tweedledum Software Package

Level 1	Level 2	Level 3
#1 Clock, calendar, queue	#1	#1
#2 Keyboard service	#2 Mass flow calculations	#2
#3	#3 Control	#3
#4 Patch	#4 Graphic trending	#4 Tank gager
#5	#5	#5
#6 Load	#6 SPYNTELL	#6 Real-time plotter
#7	#7	#7 Real-time tracker
#8 Printer buffer	#8	#8
#9	#9 Calculations	#9 Investigator (hard copy)
#10 Display build	#10 Refresh displays	#10
#11	#11 B. plotter storage	#11 Hard-copy display
#12	#12 B. tracker storage	#12 Daily report writer
#13	#13 Averaging	#13
#14	#14	#14 Batch plotter
#15	#15 Investigator (calcs)	#15 Batch tracker
#16	#16 Data storage	#16

Table 6-4
Tweedledum Variables and Constants

#000	V5	5	DEGC TERMINAL RACKS
#001	M10	14	DEGC BOTTOM REACTOR
#002	M10	14	DEGC MIDDLE REACTOR
#003	M10	14	DEGC TOP REACTOR, VAPOR SPACE
#004	M10	14	DEGC REACTOR FORWARD FLOW TO SOAKER
#005	M10	14	DEGC FIRST QUARTER SOAKER
#006	M10	14	DEGC SECOND QUARTER SOAKER
#007	M10	14	DEGC THIRD QUARTER SOAKER
#008	M10	14	DEGC FINAL QUARTER SOAKER
#009	M10	14	DEGC FLASH DRUM BOTTOM
#010	M10	14	DEGC FLASH DRUM TOP
#011	M10	14	DEGC FLASH DRUM CONDENSER LIQUID
#012	M10	14	DEGC FLASH DRUM CONDENSER VENT
#013	M10	14	DEGC NEUTRALIZER
#014	M10	14	DEGC EXTRACTOR FEED
#015	M10	14	DEGC EXTRACTOR SOLVENT
#016	M10	14	DEGC EXTRACTOR RAFFINATE
#017	M10	14	DEGC EXTRACTOR EXTRACT
#018	M10	14	DEGC EVAPORATOR OVERHEAD
#019	M10	14	DEGC EVAPORATOR CONDENSER VENT
#020	M10	14	DEGC EVAPORATOR REBOILER OUT
#021	M10	14	DEGC DISTILLATION COLUMN OVERHEAD
#022	M10	14	DEGC DISTILLATION COLUMN CONDENSER VENT
#023	M10	14	DEGC DISTILLATION COLUMN CONDENSER LIQUID
#024	M10	14	DEGC TRAY 1
#025	M10	14	DEGC TRAY 2
#026	M10	14	DEGC TRAY 3
#027	M10	14	DEGC TRAY 4
#028	M10	14	DEGC TRAY 5
#029	M10	14	DEGC TRAY 6
#030	M10	14	DEGC TRAY 7
#031	M10	14	DEGC TRAY 8
#032	M10	14	DEGC TRAY 9
#033	M10	14	DEGC TRAY 10
#034	M10	14	DEGC TRAY 11
#035	M10	14	DEGC TRAY 12
#036	M10	14	DEGC TRAY 13
#037	M10	14	DEGC TRAY 14
#038	M10	14	DEGC TRAY 15
#039	M10	14	DEGC DISTILLATION COLUMN REBOILER OUT
#040	M10	14	DEGC BUBBLE STORAGE TANK
#041	M10	14	DEGC TOIL & TROUBLE STORAGE TANK
#042	M10	14	DEGC MARES' SWEAT STORAGE TANK
#043	M10	14	DEGC TSJ TINCTURE STORAGE TANK
#044	M10	14	DEGC FROGS' SPIT STORAGE TANK

TABLE 6-4 continued

TABLE 6-4 *continued*

#045	M10	14	DEGC CRUDE STORAGE TANK
#046	M10	14	DEGC EPR RUN TANK
#047	M10	14	DEGC EPR SHIPPING TANK
#048	M10	14	DEGC WITCHES' BREW RUN TANK
#049	M10	14	DEGC WITCHES' BREW SHIPPING TANK
#050	M10	14	SPARE
#051	M10	14	SPARE
#052	M10	14	SPARE
#053	M10	14	SPARE
#054	M10	14	SPARE
#055	M10	14	SPARE
#056	M10	14	SPARE
#057	M10	14	SPARE
#058	M10	14	SPARE
#059	M10	14	SPARE
#060	M10	14	SPARE
#061	M10	14	SPARE
#062	M10	14	SPARE
#063	M10	14	SPARE
#064	V5	4	SPARE
#065	V5	4	DIVISIONS FLOW, DOUBLE FEED
#066	V5	4	DIVISIONS FLOW, BUBBLE FEED
#067	V5	4	DIVISIONS FLOW, TOIL & TROUBLE FEED
#068	V5	4	DIVISIONS FLOW, REACTOR RECYCLE
#069	V5	4	DIVISIONS FLOW, REACTOR FORWARD
#070	V5	4	DIVISIONS FLOW, REACTOR HEATER STEAM
#071	V5	4	DIVISIONS FLOW, FLASH DRUM CONDENSER VENT
#072	V5	4	DIVISIONS FLOW, FLASH DRUM CONDENSER WATER
#073	V5	4	DIVISIONS FLOW, FLASH DRUM COOLER WATER
#074	V5	4	DIVISIONS FLOW, FLASH DRUM FORWARD
#075	V5	4	DIVISIONS FLOW, MARES' SWEAT FEED
#076	V5	4	DIVISIONS FLOW, NEUTRALIZER VENT
#077	V5	3	DIVISIONS FLOW, EXTRACTORS FEED
#078	V5	3	DIVISIONS FLOW, EXTRACTORS SOLVENT
#079	V5	3	DIVISIONS FLOW, EXTRACTORS RAFFINATE
#080	V5	3	DIVISIONS FLOW, EXTRACTORS EXTRACT
#081	V5	4	DIVISIONS FLOW, EVAPORATOR RECYCLE
#082	V5	4	DIVISIONS FLOW, EVAPORATOR BOTTOMS
#083	V5	3	DIVISIONS FLOW, EVAPORATOR OVERHEAD
#084	V5	4	DIVISIONS FLOW, EVAPORATOR CONDENSER WATER
#085	V5	4	DIVISIONS FLOW, EVAPORATOR REBOILER STEAM
#086	V5	3	DIVISIONS FLOW, TSJ TINCTURE FEED
#087	V5	4	DIVISIONS FLOW, DISTILLATION COLUMN FEED
#088	V5	4	DIVISIONS FLOW, DISTILLATION COLUMN BOTTOMS
#089	V5	4	DIVISIONS FLOW, DISTILLATION COLUMN FORWARD
#090	V5	4	DIVISIONS FLOW, DISTILLATION COLUMN REFLUX
#091	V5	3	DIVISIONS FLOW, DISTILLATION COLUMN CONDENSER WATER
#092	V5	3	DIVISIONS FLOW, DISTILLATION COLUMN REBOILER STEAM

TABLE 6-4 *continued*

TABLE 6-4 continued

#093	V5	4	DIVISIONS FLOW, DISTILLATION COLUMN RECYCLE
#094	V5	4	SPARE
#095	V5	4	SPARE
#096	V5	1	LEVEL, BUBBLE STORAGE TANK
#097	V5	1	LEVEL, TOIL & TROUBLE STORAGE TANK
#098	V5	1	LEVEL, MARES' SWEAT STORAGE TANK
#099	V5	1	LEVEL, TSJ TINCTURE STORAGE TANK
#100	V5	1	LEVEL, FROGS' SPIT STORAGE TANK
#101	V5	1	LEVEL, CRUDE STORAGE TANK
#102	V5	1	LEVEL, EPR RUN TANK
#103	V5	1	LEVEL, EPR SHIPPING TANK
#104	V5	1	LEVEL, WITCHES' BREW RUN TANK
#105	V5	1	LEVEL, WITCHES' BREW SHIPPING TANK
#106	V5	24	PRESSURE, HUBBLE FEED
#107	V5	9	PRESSURE, REACTOR TOP
#108	V5	23	PRESSURE, DISTILLATION COLUMN TOP
#109	V5	23	PRESSURE, DISTILLATION COLUMN BOTTOM
#110	V5	5	DEGC HUBBLE FEED
#111	V5	2	SPARE
#112	V5	2	SPARE
#113	V5	2	SPARE
#114	V5	2	SPARE
#115	V5	2	SPARE
#116	V5	2	SPARE
#117	V5	2	SPARE
#118	V5	2	SPARE
#119	V5	2	SPARE
#120	V5	2	SPARE
#121	V5	2	SPARE
#122	V5	2	SPARE
#123	V5	2	SPARE
#124	V5	2	SPARE
#125	V5	2	SPARE
#126	V5	2	SPARE
#127	V5	2	SPARE
#128	V5	9	DISCHARGE PRESSURE, BUBBLE FEED PUMP
#129	V5	9	DISCHARGE PRESSURE, TOIL & TROUBLE FEED PUMP
#130	V5	9	DISCHARGE PRESSURE, REACTOR RECYCLE PUMP
#131	V5	7	DISCHARGE PRESSURE, MARES' SWEAT FEED PUMP
#132	V5	7	DISCHARGE PRESSURE, EXTRACTORS FEED PUMP
#133	V5	7	DISCHARGE PRESSURE, FIRST EXTRACTORS PUMP
#134	V5	7	DISCHARGE PRESSURE, SECOND EXTRACTORS PUMP
#135	V5	7	DISCHARGE PRESSURE, THIRD EXTRACTORS PUMP
#136	V5	7	DISCHARGE PRESSURE, EXTRACTORS RAFFINATE PUMP
#137	V5	7	DISCHARGE PRESSURE, EXTRACTORS SOLVENT PUMP
#138	V5	7	DISCHARGE PRESSURE, EVAPORATOR RECYCLE PUMP
#139	V5	7	DISCHARGE PRESSURE, DISTILLATION COLUMN FEED PUMP
#140	V5	7	DISCHARGE PRESSURE, TSJ TINCTURE PUMP

TABLE 6-4 continued

TABLE 6-4 continued

Tag	Type	Value	Description
#141	V5	7	DISCHARGE PRESSURE, DISTILLATION COLUMN RECYCLE PUMP
#142	V5	1	DISCHARGE PRESSURE, DISTILLATION COLUMN REFLUX PUMP
#143	V5	7	DISCHARGE PRESSURE, EPR RUN TANK PUMP
#144	V5	7	DISCHARGE PRESSURE, EPR SHIPPING TANK PUMP
#145	V5	7	DISCHARGE PRESSURE, WITCHES' BREW RUN TANK PUMP
#146	V5	7	DISCHARGE PRESSURE, WITCHES' BREW SHIPPING TANK PUMP
#147	V5	27	DISCHARGE PRESSURE, STATIC MIXER, FLASH DRUM FEED
#148	V5	27	DISCHARGE PRESSURE, STATIC MIXER, MARES' SWEAT FEED
#149	V5	27	DISCHARGE PRESSURE, STATIC MIXER, DOWNSTREAM
#150	V5	2	SPARE
#151	V5	2	SPARE
#152	V5	2	SPARE
#153	V5	2	SPARE
#154	V5	2	SPARE
#155	V5	2	SPARE
#156	V5	2	SPARE
#157	V5	2	SPARE
#158	V5	2	SPARE
#159	V5	2	SPARE
#160	V5	2	SPARE
#161	V5	2	SPARE
#162	V5	2	SPARE
#163	V5	2	SPARE
#164	V5	2	SPARE
#165	V5	2	SPARE
#166	V5	2	SPARE
#167	V5	2	SPARE
#168	V5	2	SPARE
#169	V5	2	SPARE
#170	V5	2	SPARE
#171	V5	2	SPARE
#172	V5	2	SPARE
#173	V5	2	SPARE
#174	V5	2	SPARE
#175	V5	2	SPARE
#176	V5	2	SPARE
#177	V5	2	SPARE
#178	V5	2	SPARE
#179	V5	2	SPARE
#180	V5	2	SPARE
#181	V5	2	SPARE
#182	V5	2	SPARE
#183	V5	2	SPARE
#184	V5	2	SPARE
#185	V5	2	SPARE
#186	V5	2	SPARE
#187	V5	2	SPARE
#188	V5	2	SPARE
#189	V5	2	SPARE
#190	V5	2	SPARE
#191	V5	2	SPARE
#192	V5	2	SPARE
#193	V5	2	SPARE
#194	V5	2	SPARE
#195	V5	2	SPARE
#196	V5	2	SPARE
#197	V5	2	SPARE
#198	V5	2	SPARE
#199	V5	2	SPARE
#200	V5	2	SPARE
#201	V5	2	SPARE
#202	V5	2	SPARE
#203	V5	2	SPARE
#204	V5	2	SPARE
#205	V5	2	SPARE
#206	V5	2	SPARE
#207	V5	2	SPARE
#208	V5	2	SPARE
#209	V5	2	SPARE
#210	V5	2	SPARE
#211	V5	2	SPARE
#212	V5	2	SPARE

TABLE 6-4 continued

TABLE 6-4 continued

#213	V5	2	SPARE
#214	V5	2	SPARE
#215	V5	2	SPARE
#216	V5	2	SPARE
#217	V5	2	SPARE
#218	V5	2	SPARE
#219	V5	2	SPARE
#220	V5	2	SPARE
#221	V5	2	SPARE
#222	V5	2	SPARE
#223	V5	2	SPARE
#224	V5	2	REACTOR LEVEL
#225	V5	9	REACTOR RECYCLE TEMPERATURE
#226	V5	9	SOAKER PRESSURE
#227	V5	27	FLASH DRUM CONDENSER PRESSURE
#228	V5	5	FLASH DRUM CONDENSER LIQUID TEMPERATURE
#229	V5	5	FLASH DRUM FORWARD FLOW TEMPERATURE
#230	V5	2	FLASH DRUM LEVEL
#231	V5	8	NEUTRALIZER PH
#232	V5	2	NEUTRALIZER LEVEL
#233	V5	2	FIRST EXTRACTORS DRUM LEVEL
#234	V5	2	SECOND EXTRACTORS DRUM LEVEL
#235	V5	2	THIRD EXTRACTORS DRUM LEVEL
#236	V5	7	FIRST EXTRACTORS DRUM PRESSURE
#237	V5	2	EVAPORATOR LEVEL
#238	V5	2	EVAPORATOR BOTTOMS TEMPERATURE
#239	V5	7	EVAPORATOR CONDENSER LIQUID TEMPERATURE
#240	V5	2	EVAPORATOR CONDENSER LIQUID LEVEL
#241	V5	2	DISTILLATION COLUMN BOTTOMS LEVEL
#242	V5	2	DISTILLATION COLUMN REFLUX DRUM LEVEL
#243	V5	7	DISTILLATION COLUMN CONDENSER LIQUID TEMPERATURE
#244	V5	9	DISTILLATION COLUMN BOTTOMS TEMPERATURE
#245	V5	1	SPARE
#246	V5	1	SPARE
#247	V5	1	SPARE
#248	V5	1	SPARE
#249	V5	1	SPARE
#250	V5	1	SPARE
#251	V5	1	SPARE
#252	V5	1	SPARE
#253	V5	1	SPARE
#254	V5	1	SPARE
#255	V5	1	SPARE
#256	V5	2	DOUBLE FEED CONTROLLER SET-POINT
#257	V5	2	BUBBLE FEED CONTROLLER SET-POINT
#258	V5	2	TOIL & TROUBLE FEED CONTROLLER SET-POINT
#259	V5	2	REACTOR RECYCLE FLOW CONTROLLER SET-POINT
#260	V5	2	REACTOR LEVEL CONTROLLER SET-POINT

TABLE 6-4 continued

TABLE 6-4 continued

#261	V5	2	REACTOR TEMPERATURE CONTROLLER SET-POINT
#262	V5	2	SOAKER PRESSURE CONTROLLER SET-POINT
#263	V5	2	FLASH DRUM PRESSURE CONTROLLER SET-POINT
#264	V5	2	FLASH DRUM CONDENSER TEMPERATURE CONTROLLER SET-POINT
#265	V5	2	FLASH DRUM COOLER TEMPERATURE CONTROLLER SET-POINT
#266	V5	2	FLASH DRUM LEVEL CONTROLLER SET-POINT
#267	V5	2	NEUTRALIZER PH CONTROLLER SET-POINT
#268	V5	2	NEUTRALISER LEVEL CONTROLLER SET-POINT
#269	V5	2	FIRST EXTRACTOR DRUM LEVEL CONTROLLER SET-POINT
#270	V5	2	SECOND EXTRACTOR DRUM LEVEL CONTROLLER SET-POINT
#271	V5	2	THIRD EXTRACTOR DRUM LEVEL CONTROLLER SET-POINT
#272	V5	2	EXTRACTOR PRESSURE CONTROLLER SET-POINT
#273	V5	2	SOLVENT FEED FLOW CONTROLLER SET-POINT
#274	V5	2	EVAPORATOR LEVEL CONTROLLER SET-POINT
#275	V5	2	EVAPORATOR TEMPERATURE CONTROLLER SET-POINT
#276	V5	2	EVAPORATOR CONDENSER TEMPERATURE CONTROLLER SET-POINT
#277	V5	2	EVAPORATOR CONDENSER LEVEL CONTROLLER SET-POINT
#278	V5	2	TSJ TINCTURE FEED FLOW CONTROLLER SET-POINT
#279	V5	2	DISTILLATION COLUMN FEED FLOW CONTROLLER SET-POINT
#280	V5	2	DISTILLATION COLUMN LEVEL CONTROLLER SET-POINT
#281	V5	2	DISTILLATION COLUMN REFLUX DRUM LEVEL CONTROLLER SET-POINT
#282	V5	2	DISTILLATION COLUMN REFLUX FLOW CONTROLLER SET-POINT
#283	V5	2	DISTILLATION COLUMN CONDENSER TEMPERATURE CONTROLLER SET-POINT
#284	V5	2	DISTILLATION COLUMN BOTTOMS TEMPERATURE CONTROLLER SET-POINT
#285	V5	2	SPARE
#286	V5	2	SPARE
#287	V5	2	SPARE
#288	V5	1	DOUBLE FEED CONTROL VALVE POSITION
#289	V5	1	BUBBLE FEED CONTROL VALVE POSITION
#290	V5	1	TOIL & TROUBLE FEED CONTROL VALVE POSITION
#291	V5	1	REACTOR RECYCLE CONTROL VALVE POSITION
#292	V5	1	REACTOR FORWARD FLOW CONTROL VALVE POSITION
#293	V5	1	REACTOR HEATER STEAM FLOW CONTROL VALVE POSITION
#294	V5	1	SOAKER FORWARD FLOW CONTROL VALVE POSITION
#295	V5	1	FLASH DRUM VENT CONTROL VALVE POSITION
#296	V5	1	FLASH DRUM CONDENSER WATER CONTROL VALVE POSITION
#297	V5	1	FLASH DRUM COOLER WATER CONTROL VALVE POSITION
#298	V5	1	FLASH DRUM FORWARD FLOW CONTROL VALVE POSITION
#299	V5	1	MARES' SWEAT FEED CONTROL VALVE POSITION
#300	V5	1	EXTRACTORS FEED CONTROL VALVE POSITION
#301	V5	1	FIRST EXTRACTORS DRUM CONTROL VALVE POSITION
#302	V5	1	SECOND EXTRACTORS DRUM CONTROL VALVE POSITION
#303	V5	1	THIRD EXTRACTORS DRUM CONTROL VALVE POSITION
#304	V5	1	EXTRACT CONTROL VALVE POSITION
#305	V5	1	SOLVENT CONTROL VALVE POSITION
#306	V5	1	EVAPORATOR BOTTOMS CONTROL VALVE POSITION
#307	V5	1	EVAPORATOR REBOILER STEAM CONTROL VALVE POSITION
#308	V5	1	EVAPORATOR CONDENSER WATER CONTROL VALVE POSITION

TABLE 6-4 continued

TABLE 6-4 continued

#309	V5	1	EVAPORATOR FORWARD FLOW CONTROL VALVE POSITION
#310	V5	1	TSJ TINCTURE FEED CONTROL VALVE POSITION
#311	V5	1	DISTILLATION COLUMN FEED CONTROL VALVE POSITION
#312	V5	1	DISTILLATION COLUMN BOTTOMS FLOW CONTROL VALVE POSITION
#313	V5	1	DISTILLATION COLUMN FORWARD FLOW CONTROL VALVE POSITION
#314	V5	1	DISTILLATION COLUMN REFLUX CONTROL VALVE POSITION
#315	V5	1	DISTILLATION COLUMN CONDENSER WATER CONTROL VALVE POSITION
#316	V5	1	DISTILLATION COLUMN REBOILER STEAM CONTROL VALVE POSITION
#317	V5	1	SPARE
#318	V5	1	SPARE
#319	V5	1	SPARE
#320			MAX RATE BUBBLE MASS FLOW (REACTION STEP)
#321			MIN RATE BUBBLE MASS FLOW (REACTION STEP)
#322			MAX RATE EXTRACTOR FEED (EXTRACTION STEP)
#323			MIN RATE EXTRACTOR FEED (EXTRACTION STEP)
#324			MAX RATE DISTILLATION COLUMN FEED (DISTILLATION STEP)
#325			MIN RATE DISTILLATION COLUMN FEED (DISTILLATION STEP)
#326			MASS FLOW TOIL IN FEED
#327			MASS FLOW TROUBLE IN FEED
#328			VOLUME FLOW BUBBLE
#329			VOLUME FLOW TOIL & TROUBLE
#330			VOLUME FLOW WET CRUDE
#331			MASS FLOW EPR (WET CRUDE)
#332			MASS FLOW WITCHES BREW (WET CRUDE)
#333			MASS FLOW TROUBLE (WET CRUDE)
#334			MASS FLOW HEAVIES (WET CRUDE)
#335			VOLUME FLOW RAFFINATE
#336			MASS FLOW TROUBLE (RAFFINATE)
#337			MASS FLOW HEAVIES (RAFFINATE)
#338			MASS FLOW WATER (BUBBLE)
#339			MASS FLOW WATER (TOIL & TROUBLE)
#340			MASS FLOW WATER (WET CRUDE)
#341			MASS FLOW WATER (RAFFINATE)
#342			VOLUME FLOW EXTRACT
#343			MASS FLOW EPR (EXTRACT)
#344			MASS FLOW WITCHES BREW (EXTRACT)
#345			MASS FLOW FROG SPIT (EXTRACT)
#346			VOLUME FLOW DISTILLATTION COLUMN FEED
#347			MASS FLOW WITCHES BREW (DISCOL FEED)
#348			MASS FLOW EPR (DISCOL FEED)
#349			MASS FLOW WITCHES BREW TO NEUTRALIZER
#350			MASS FLOW EPR TO NEUTRALIZER
#351			SPARE
#352			ANALYSIS BUBBLE IN BUBBLE FEED
#353			ANALYSIS TOIL IN T&T FEED
#354			ANALYSIS TROUBLE IN T&T FEED

TABLE 6-4 continued

TABLE 6-4 continued

#355	ANALYSIS DOUBLE IN DOUBLE FEED
#356	ANALYSIS NOXIOUS FLASH DRUM VENT
#357	ANALYSIS GAZ NEUTRALIZER VENT
#358	ANALYSIS EPR IN WET CRUDE
#359	ANALYSIS WITCHES BREW IN WET CRUDE
#360	ANALYSIS TROUBLE IN WET CRUDE
#361	ANALYSIS HEAVIES WET CRUDE
#362	ANALYSIS TROUBLE IN RAFFINATE
#363	ANALYSIS HEAVIES IN RAFFINATE
#364	ANALYSIS EPR IN EXTRACT
#365	ANALYSIS WITCHES BREW IN EXTRACT
#366	ANALYSIS FROG SPIT IN EXTRACT
#367	ANALYSIS WITCHES BREW IN DISCOL FEED
#368	ANALYSIS EPR IN DISCOL FEED
#369	ANALYSIS WITCHES BREW IN TOP PRODUCT
#370	ANALYSIS EPR IN TOP PRODUCT
#371	ANALYSIS ALDEHYDE IN TOP PRODUCT
#372	ANALYSIS WITCHES BREW IN BTM PRODUCT
#373	ANALYSIS EPR IN BTM PRODUCT
#374	ANALYSIS WITCHES BREW IN WB RUN TANK
#375	ANALYSIS EPR IN WB RUN TANK
#376	ANALYSIS WITCHES BREW IN WB SHIP TANK
#377	ANALYSIS EPR IN WB SHIP TANK
#378	ANALYSIS WITCHES BREW IN EPR RUN TANK
#379	ANALYSIS EPR IN EPR RUN TANK
#380	ANALYSIS ALDEHYDE IN EPR RUN TANK
#381	ANALYSIS WITCHES BREW IN EPR SHIP TANK
#382	ANALYSIS EPR IN EPR SHIP TANK
#383	ANALYSIS ALDEHYDE IN EPR SHIP TANK
#384	SPARE
#385	SPARE
#386	SPARE
#387	SPARE
#388	SPARE
#389	SPARE
#390	SPARE
#391	SPARE
#392	SPARE
#393	SPARE
#394	SPARE
#395	SPARE
#396	SPARE
#397	SPARE
#398	SPARE
#399	SPARE
#400	SPARE

TABLE 6-4 continued

TABLE 6-4 continued

#401	SPARE	#424	SPARE	#447	SPARE
#402	SPARE	#425	SPARE	#448	SPARE
#403	SPARE	#426	SPARE	#449	SPARE
#404	SPARE	#427	SPARE	#450	SPARE
#405	SPARE	#428	SPARE	#451	SPARE
#406	SPARE	#429	SPARE	#452	SPARE
#407	SPARE	#430	SPARE	#453	SPARE
#408	SPARE	#431	SPARE	#454	SPARE
#409	SPARE	#432	SPARE	#455	SPARE
#410	SPARE	#433	SPARE	#456	SPARE
#411	SPARE	#434	SPARE	#457	SPARE
#412	SPARE	#435	SPARE	#458	SPARE
#413	SPARE	#436	SPARE	#459	SPARE
#414	SPARE	#437	SPARE	#460	SPARE
#415	SPARE	#438	SPARE	#461	SPARE
#416	SPARE	#439	SPARE	#462	SPARE
#417	SPARE	#440	SPARE	#463	SPARE
#418	SPARE	#441	SPARE	#464	SPARE
#419	SPARE	#442	SPARE	#465	SPARE
#420	SPARE	#443	SPARE	#466	SPARE
#421	SPARE	#444	SPARE	#467	SPARE
#422	SPARE	#445	SPARE	#468	SPARE
#423	SPARE	#446	SPARE	#469	SPARE

TABLE 6-4 continued

TABLE 6-4 continued

#	Description
#470	SPARE
#471	SPARE
#472	SPARE
#473	SPARE
#474	SPARE
#475	SPARE
#476	SPARE
#477	SPARE
#478	SPARE
#479	SPARE
#480	DENSITY OF BUBBLE, BUBBLE STORAGE TANK
#481	DENSITY OF TOIL & TROUBLE, TOIL & TROUBLE STORAGE TANK
#482	DENSITY OF MARES' SWEAT, MARES' SWEAT STORAGE TANK
#483	DENSITY OF TSJ TINCTURE, TSJ TINCTURE STORAGE TANK
#484	DENSITY OF FROGS' SPIT, FROGS' SPIT STORAGE TANK
#485	DENSITY OF CRUDE, CRUDE STORAGE TANK
#486	DENSITY OF EFR, EFR RUN TANK
#487	DENSITY OF WITCHES' BREW, WITCHES' BREW RUN TANK
#488	DENSITY OF EFR, EFR SHIPPING TANK
#489	DENSITY OF WITCHES' BREW, WITCHES' BREW SHIPPING TANK
#490	OPENING INVENTORY BUBBLE
#491	OPENING INVENTORY TOIL & TROUBLE
#492	OPENING INVENTORY MARES' SWEAT
#493	OPENING INVENTORY TSJ TINCTURE
#494	OPENING INVENTORY FROGS' SPIT
#495	OPENING INVENTORY CRUDE
#496	OPENING INVENTORY EFR, RUN TANK
#497	OPENING INVENTORY EFR, SHIPPING TANK
#498	OPENING INVENTORY WITCHES' BREW, RUN TANK
#499	OPENING INVENTORY WITCHES' BREW, SHIPPING TANK
#500	CURRENT INVENTORY BUBBLE
#501	CURRENT INVENTORY TOIL & TROUBLE
#502	CURRENT INVENTORY MARES' SWEAT
#503	CURRENT INVENTORY TSJ TINCTURE
#504	CURRENT INVENTORY FROGS' SPIT
#505	CURRENT INVENTORY CRUDE
#506	CURRENT INVENTORY EFR, RUN TANK
#507	CURRENT INVENTORY EFR, SHIPPING TANK
#508	CURRENT INVENTORY WITCHES' BREW, RUN TANK
#509	CURRENT INVENTORY WITCHES' BREW, SHIPPING TANK
#510	SPARE
#511	SPARE
#512	DIAMETER BUBBLE STORAGE TANK
#513	HEIGHT BUBBLE STORAGE TANK
#514	DIAMETER TOIL & TROUBLE STORAGE TANK
#515	HEIGHT TOIL & TROUBLE STORAGE TANK
#516	DIAMETER MARES' SWEAT STORAGE TANK
#517	HEIGHT MARES' SWEAT STORAGE TANK

TABLE 6-4 continued

TABLE 6-4 continued

#	Description
#518	DIAMETER TSJ TINCTURE STORAGE TANK
#519	HEIGHT TSJ TINCTURE STORAGE TANK
#520	DIAMETER FROGS' SPIT STORAGE TANK
#521	HEIGHT FROGS' SPIT STORAGE TANK
#522	DIAMETER CRUDE STORAGE TANK
#523	HEIGHT CRUDE STORAGE TANK
#524	DIAMETER EFR RUN TANK
#525	HEIGHT EFR RUN TANK
#526	DIAMETER WITCHES' BREW RUN TANK
#527	HEIGHT WITCHES' BREW RUN TANK
#528	DIAMETER EFR SHIPPING TANK
#529	HEIGHT EFR SHIPPING TANK
#530	DIAMETER WITCHES' BREW SHIPPING TANK
#531	HEIGHT WITCHES' BREW SHIPPING TANK
#532	SPARE
#533	SPARE
#534	SPARE
#535	SPARE
#536	SPARE
#537	SPARE
#538	SPARE
#539	SPARE
#540	SPARE
#541	SPARE
#542	SPARE
#543	SPARE
#544	SPARE
#545	MASS FLOW DOUBLE FEED
#546	MASS FLOW BUBBLE FEED
#547	MASS FLOW TOIL & TROUBLE FEED
#548	MASS FLOW REACTOR RECYCLE
#549	MASS FLOW REACTOR FORWARD
#550	MASS FLOW REACTOR HEATER STEAM
#551	MASS FLOW FLASH DRUM CONDENSER VENT
#552	MASS FLOW FLASH DRUM CONDENSER WATER
#553	MASS FLOW FLASH DRUM COOLER WATER
#554	MASS FLOW FLASH DRUM FORWARD
#555	MASS FLOW MARES' SWEAT FEED
#556	MASS FLOW NEUTRALIZER VENT
#557	MASS FLOW EXTRACTORS FEED
#558	MASS FLOW EXTRACTORS SOLVENT
#559	MASS FLOW EXTRACTORS RAFFINATE
#560	MASS FLOW EXTRACTORS EXTRACT
#561	MASS FLOW EVAPORATOR RECYCLE
#562	MASS FLOW EVAPORATOR BOTTOMS
#563	MASS FLOW EVAPORATOR OVERHEAD
#564	MASS FLOW EVAPORATOR CONDENSER WATER
#565	MASS FLOW EVAPORATOR REBOILER STEAM

TABLE 6-4 continued

TABLE 6-4 continued

#566	MASS FLOW TSJ TINCTURE FEED
#567	MASS FLOW DISTILLATION COLUMN FEED
#568	MASS FLOW DISTILLATION COLUMN BOTTOMS
#569	MASS FLOW DISTILLATION COLUMN FORWARD
#570	VOLUME FLOW DISTILLATION COLUMN REFLUX
#571	VOLUME FLOW DISTILLATION COLUMN CONDENSER WATER
#572	MASS FLOW DISTILLATION COLUMN REBOILER STEAM
#573	MASS FLOW DISTILLATION COLUMN RECYCLE
#574	SPARE
#575	SPARE
#576	SPARE
#577	ORIFICE PLATE FACTOR DOUBLE FEED
#578	ORIFICE PLATE FACTOR BUBBLE FEED
#579	ORIFICE PLATE FACTOR TOIL & TROUBLE FEED
#580	ORIFICE PLATE FACTOR REACTOR RECYCLE
#581	ORIFICE PLATE FACTOR REACTOR FORWARD
#582	ORIFICE PLATE FACTOR REACTOR HEATER STEAM
#583	ORIFICE PLATE FACTOR FLASH DRUM CONDENSER VENT
#584	ORIFICE PLATE FACTOR FLASH DRUM CONDENSER WATER
#585	ORIFICE PLATE FACTOR FLASH DRUM COOLER WATER
#586	ORIFICE PLATE FACTOR FLASH DRUM FORWARD
#587	ORIFICE PLATE FACTOR MARES' SWEAT FEED
#588	ORIFICE PLATE FACTOR NEUTRALIZER VENT
#589	ORIFICE PLATE FACTOR EXTRACTORS FEED
#590	ORIFICE PLATE FACTOR EXTRACTORS SOLVENT
#591	ORIFICE PLATE FACTOR EXTRACTORS RAFFINATE
#592	ORIFICE PLATE FACTOR EXTRACTORS EXTRACT
#593	ORIFICE PLATE FACTOR EVAPORATOR RECYCLE
#594	ORIFICE PLATE FACTOR EVAPORATOR BOTTOMS
#595	ORIFICE PLATE FACTOR EVAPORATOR OVERHEAD
#596	ORIFICE PLATE FACTOR EVAPORATOR CONDENSER WATER
#597	ORIFICE PLATE FACTOR EVAPORATOR REBOILER STEAM
#598	ORIFICE PLATE FACTOR TSJ TINCTURE FEED
#599	ORIFICE PLATE FACTOR DISTILLATION COLUMN FEED
#600	ORIFICE PLATE FACTOR DISTILLATION COLUMN BOTTOMS
#601	ORIFICE PLATE FACTOR DISTILLATION COLUMN FORWARD
#602	ORIFICE PLATE FACTOR DISTILLATION COLUMN REFLUX
#603	ORIFICE PLATE FACTOR DISTILLATION COLUMN CONDENSER WATER
#604	ORIFICE PLATE FACTOR DISTILLATION COLUMN REBOILER STEAM
#605	ORIFICE PLATE FACTOR DISTILLATION COLUMN RECYCLE
#606	SPARE
#607	SPARE
#608	SPARE
#609	DAY AVERAGE MASS FLOW, DOUBLE FEED
#610	DAY AVERAGE MASS FLOW, BUBBLE FEED
#611	DAY AVERAGE MASS FLOW, TOIL & TROUBLE FEED
#612	DAY AVERAGE MASS FLOW, REACTOR RECYCLE

TABLE 6-4 continued

TABLE 6-4 continued

#	Description
#613	DAY AVERAGE MASS FLOW, REACTOR FORWARD
#614	DAY AVERAGE MASS FLOW, REACTOR HEATER STEAM
#615	DAY AVERAGE MASS FLOW, FLASH DRUM CONDENSER VENT
#616	DAY AVERAGE MASS FLOW, FLASH DRUM CONDENSER WATER
#617	DAY AVERAGE MASS FLOW, FLASH DRUM COOLER WATER
#618	DAY AVERAGE MASS FLOW, FLASH DRUM FORWARD
#619	DAY AVERAGE MASS FLOW, MARES' SWEAT FEED
#620	DAY AVERAGE MASS FLOW, NEUTRALIZER VENT
#621	DAY AVERAGE MASS FLOW, EXTRACTORS FEED
#622	DAY AVERAGE MASS FLOW, EXTRACTORS SOLVENT
#623	DAY AVERAGE MASS FLOW, EXTRACTORS RAFFINATE
#624	DAY AVERAGE MASS FLOW, EXTRACTORS EXTRACT
#625	DAY AVERAGE MASS FLOW, EVAPORATOR RECYCLE
#626	DAY AVERAGE MASS FLOW, EVAPORATOR BOTTOMS
#627	DAY AVERAGE MASS FLOW, EVAPORATOR OVERHEAD
#628	DAY AVERAGE MASS FLOW, EVAPORATOR CONDENSER WATER
#629	DAY AVERAGE MASS FLOW, EVAPORATOR REBOILER STEAM
#630	DAY AVERAGE MASS FLOW, TSJ TINCTURE FEED
#631	DAY AVERAGE MASS FLOW, DISTILLATION COLUMN FEED
#632	DAY AVERAGE MASS FLOW, DISTILLATION COLUMN BOTTOMS
#633	DAY AVERAGE MASS FLOW, DISTILLATION COLUMN FORWARD
#634	DAY AVERAGE MASS FLOW, DISTILLATION COLUMN REFLUX
#635	DAY AVERAGE MASS FLOW, DISTILLATION COLUMN CONDENSER WATER
#636	DAY AVERAGE MASS FLOW, DISTILLATION COLUMN REBOILER STEAM
#637	DAY AVERAGE MASS FLOW, DISTILLATION COLUMN RECYCLE
#638	SPARE
#639	SPARE
#640	DOUBLE MODE
#641	DOUBLE FLAG
#642	DOUBLE MASS FLOW TARGET
#643	DOUBLE RATIO HISTORY
#644	DOUBLE TARGET RATIO
#645	DOUBLE CURRENT RATIO
#646	DOUBLE CVP AT FIRST FLOW (VALVE OFFSET)
#647	DOUBLE FLOW AT ZERO CVP (FLOW OFFSET)
#648	DOUBLE TARGET VALVE POSITION
#649	DOUBLE RAMP DELTA FOR ADJUSTED RATIO
#650	DOUBLE MAXIMUM SINGLE CYCLE MASS FLOW ADJUSTMENT
#651	DOUBLE SPARE
#652	BUBBLE MODE
#653	BUBBLE FLAG
#654	BUBBLE LONG TERM MASS FLOW TARGET
#655	BUBBLE SHORT TERM MASS FLOW TARGET
#656	BUBBLE TARGET HISTORY
#657	BUBBLE CVP AT FIRST FLOW (VALVE OFFSET)
#658	BUBBLE FLOW AT ZERO CVP (FLOW OFFSET)
#659	BUBBLE TARGET VALVE POSITION

TABLE 6-4 continued

TABLE 6-4 continued

#660 BUBBLE FAST RAMP FLOW LIMITS
#661 BUBBLE FAST RAMP DELTA
#662 BUBBLE SLOW RAMP FLOW LIMITS
#663 BUBBLE SLOW RAMP DELTA
#664 BUBBLE SPARE
#665 BUBBLE SPARE
#666 TOIL & TROUBLE MODE
#667 TOIL & TROUBLE FLAG
#668 TOIL & TROUBLE MASS FLOW TARGET
#669 TOIL & TROUBLE RATIO HISTORY
#670 TOIL & TROUBLE TARGET RATIO
#671 TOIL & TROUBLE CURRENT RATIO
#672 TOIL & TROUBLE CVP AT FIRST FLOW (VALVE OFFSET)
#673 TOIL & TROUBLE FLOW AT ZERO CVP (FLOW OFFSET)
#674 TOIL & TROUBLE TARGET VALVE POSITION
#675 TOIL & TROUBLE RAMP DELTA FOR ADJUSTED RATIO
#676 TOIL & TROUBLE MAXIMUM SINGLE CYCLE MASS FLOW ADJUSTMENT
#677 TOIL & TROUBLE SPARE
#678 REACTOR RECYCLE MODE
#679 REACTOR RECYCLE FLAG
#680 REACTOR RECYCLE LONG TERM MASS FLOW TARGET
#681 REACTOR RECYCLE SHORT TERM MASS FLOW TARGET

#682 REACTOR RECYCLE TARGET HISTORY
#683 REACTOR RECYCLE CVP AT FIRST FLOW (VALVE OFFSET)
#684 REACTOR RECYCLE FLOW AT ZERO CVP (FLOW OFFSET)
#685 REACTOR RECYCLE TARGET VALVE POSITION
#686 REACTOR RECYCLE FAST RAMP FLOW LIMITS
#687 REACTOR RECYCLE FAST RAMP DELTA
#688 REACTOR RECYCLE SLOW RAMP FLOW LIMITS
#689 REACTOR RECYCLE SLOW RAMP DELTA
#690 REACTOR RECYCLE SPARE
#691 REACTOR RECYCLE SPARE
#692 REACTOR LEVEL MODE
#693 REACTOR LEVEL FLAG
#694 REACTOR LEVEL MASS FLOW TARGET
#695 REACTOR LEVEL RATIO HISTORY
#696 REACTOR LEVEL CURRENT RATIO
#697 REACTOR LEVEL CVP AT FIRST FLOW (VALVE OFFSET)
#698 REACTOR LEVEL FLOW AT ZERO CVP (FLOW OFFSET)
#699 REACTOR LEVEL TARGET VALVE POSITION
#700 REACTOR LEVEL HISTORY
#701 REACTOR LEVEL MAXIMUM SINGLE CYCLE MASS FLOW ADJUSTMENT
#702 REACTOR LEVEL LOW LIMIT
#703 REACTOR LEVEL HIGH LIMIT

TABLE 6-4 continued

TABLE 6-4 *continued*

TABLE 6-4 *continued*

```
#704 REACTOR LEVEL: MULTIPLIER, LEVEL DEVIATION TO MASS FLOW ADJUSTMENT
#705 REACTOR LEVEL CVP TO CVP MULTIPLIER
#706 REACTOR LEVEL SPARE
#707 REACTOR TEMPERATURE MODE
#708 REACTOR TEMPERATURE FLAG
#709 REACTOR TEMPERATURE MASS FLOW TARGET
#710 REACTOR TEMPERATURE RATIO HISTORY
#711 REACTOR TEMPERATURE CURRENT RATIO
#712 REACTOR TEMPERATURE CVP AT FIRST FLOW (VALVE OFFSET)
#713 REACTOR TEMPERATURE FLOW AT ZERO CVP (FLOW OFFSET)
#714 REACTOR TEMPERATURE TARGET VALVE POSITION
#715 REACTOR TEMPERATURE MAXIMUM SINGLE CYCLE MASS FLOW ADJUSTMENT
#716 REACTOR TEMPERATURE LOW LIMIT
#717 REACTOR TEMPERATURE HIGH LIMIT
#718 REACTOR TEMPERATURE OUT OF LIMIT SUCCESSIVE COUNTER
#719 REACTOR TEMPERATURE OUT OF LIMIT MAX COUNT
#720 REACTOR TEMPERATURE DELAY COUNTER
#721 REACTOR TEMPERATURE DELAY COUNT
#722 REACTOR TEMPERATURE: MULTIPLIER, TEMPERATURE DEVIATION TO MASS FLOW ADJUSTMENT
#723 REACTOR TEMPERATURE SPARE
#724 FLASH DRUM CONDENSER TEMPERATURE MODE
#725 FLASH DRUM CONDENSER TEMPERATURE FLAG
#726 FLASH DRUM CONDENSER TEMPERATURE MASS FLOW TARGET
```

TABLE 6-4 continued

```
#727 FLASH DRUM CONDENSER TEMPERATURE RATIO HISTORY
#728 FLASH DRUM CONDENSER TEMPERATURE CURRENT RATIO
#729 FLASH DRUM CONDENSER TEMPERATURE CVP AT FIRST FLOW (VALVE OFFSET)
#730 FLASH DRUM CONDENSER TEMPERATURE FLOW AT ZERO CVP (FLOW OFFSET)
#731 FLASH DRUM CONDENSER TEMPERATURE TARGET VALVE POSITION
#732 FLASH DRUM CONDENSER TEMPERATURE MAXIMUM SINGLE CYCLE MASS FLOW ADJUSTMENT
#733 FLASH DRUM CONDENSER TEMPERATURE LOW LIMIT
#734 FLASH DRUM CONDENSER TEMPERATURE HIGH LIMIT
#735 FLASH DRUM CONDENSER TEMPERATURE OUT OF LIMIT SUCCESSIVE COUNTER
#736 FLASH DRUM CONDENSER TEMPERATURE OUT OF LIMIT MAX COUNT
#737 FLASH DRUM CONDENSER TEMPERATURE DELAY COUNTER
#738 FLASH DRUM CONDENSER TEMPERATURE DELAY COUNT
#739 FLASH DRUM CONDENSER TEMPERATURE: MULTIPLIER, TEMPERATURE DEVIATION TO MASS FLOW ADJUSTMENT
#740 FLASH DRUM CONDENSER TEMPERATURE SPARE
#741 FLASH DRUM COOLER TEMPERATURE MODE
#742 FLASH DRUM COOLER TEMPERATURE FLAG
#743 FLASH DRUM COOLER TEMPERATURE MASS FLOW TARGET
#744 FLASH DRUM COOLER TEMPERATURE RATIO HISTORY
#745 FLASH DRUM COOLER TEMPERATURE CURRENT RATIO
#746 FLASH DRUM COOLER TEMPERATURE CVP AT FIRST FLOW (VALVE OFFSET)
#747 FLASH DRUM COOLER TEMPERATURE FLOW AT ZERO CVP (FLOW OFFSET)
#748 FLASH DRUM COOLER TEMPERATURE TARGET VALVE POSITION
#749 FLASH DRUM COOLER TEMPERATURE MAXIMUM SINGLE CYCLE MASS FLOW ADJUSTMENT
```

TABLE 6-4 continued

TABLE 6-4 continued

TABLE 6-4 continued

```
#750 FLASH DRUM COOLER TEMPERATURE LOW LIMIT
#751 FLASH DRUM COOLER TEMPERATURE HIGH LIMIT
#752 FLASH DRUM COOLER TEMPERATURE OUT OF LIMIT SUCCESSIVE COUNTER
#753 FLASH DRUM COOLER TEMPERATURE OUT OF LIMIT MAX COUNT
#754 FLASH DRUM COOLER TEMPERATURE DELAY COUNTER
#755 FLASH DRUM COOLER TEMPERATURE DELAY COUNT
#756 FLASH DRUM COOLER TEMPERATURE: MULTIPLIER, TEMPERATURE DEVIATION TO MASS FLOW ADJUSTMENT
#757 FLASH DRUM COOLER TEMPERATURE SPARE
#758 FLASH DRUM LEVEL MODE
#759 FLASH DRUM LEVEL FLAG
#760 FLASH DRUM LEVEL MASS FLOW TARGET
#761 FLASH DRUM LEVEL RATIO HISTORY
#762 FLASH DRUM LEVEL CURRENT RATIO
#763 FLASH DRUM LEVEL CVP AT FIRST FLOW (VALVE OFFSET)
#764 FLASH DRUM LEVEL FLOW AT ZERO CVP (FLOW OFFSET)
#765 FLASH DRUM LEVEL TARGET VALVE POSITION
#766 FLASH DRUM LEVEL HISTORY
#767 FLASH DRUM LEVEL MAXIMUM SINGLE CYCLE MASS FLOW ADJUSTMENT
#768 FLASH DRUM LEVEL LOW LIMIT
#769 FLASH DRUM LEVEL HIGH LIMIT
#770 FLASH DRUM LEVEL: MULTIPLIER. LEVEL DEVIATION TO MASS FLOW ADJUSTMENT
#771 FLASH DRUM LEVEL CVP TO CVP MULTIPLIER
#772 FLASH DRUM LEVEL SPARE
```

TABLE 6-4 continued

TABLE 6-4 continued

```
#773  NEUTRALIZER PH MODE
#774  NEUTRALIZER PH FLAG
#775  NEUTRALIZER PH MASS FLOW TARGET
#776  NEUTRALIZER PH RATIO HISTORY
#777  NEUTRALIZER PH CURRENT RATIO
#778  NEUTRALIZER PH MINIMUM RATIO
#779  NEUTRALIZER PH CVP AT FIRST FLOW (VALVE OFFSET)
#780  NEUTRALIZER PH FLOW AT ZERO CVP (FLOW OFFSET)
#781  NEUTRALIZER PH TARGET VALVE POSITION
#782  NEUTRALIZER PH MINIMUM FLOW
#783  NEUTRALIZER PH MAXIMUM SINGLE CYCLE MASS FLOW ADJUSTMENT
#784  NEUTRALIZER PH LOW LIMIT
#785  NEUTRALIZER PH HIGH LIMIT
#786  NEUTRALIZER PH OUT OF LIMIT SUCCESSIVE COUNTER
#787  NEUTRALIZER PH OUT OF LIMIT MAX COUNT
#788  NEUTRALIZER PH DELAY COUNTER
#789  NEUTRALIZER PH DELAY COUNT
#790  NEUTRALIZER PH: MULTIPLIER, PH DEVIATION TO MASS FLOW ADJUSTMENT
#791  NEUTRALIZER PH SPARE
#792  EXTRACTOR FEED MODE
#793  EXTRACTOR FEED FLAG
#794  EXTRACTOR FEED LONG TERM MASS FLOW TARGET
#795  EXTRACTOR FEED SHORT TERM MASS FLOW TARGET
```

TABLE 6-4 continued

TABLE 6-4 continued

```
#796  EXTRACTOR FEED TARGET HISTORY
#797  EXTRACTOR FEED CVP AT FIRST FLOW (VALVE OFFSET)
#798  EXTRACTOR FEED FLOW AT ZERO CVP (FLOW OFFSET)
#799  EXTRACTOR FEED TARGET VALVE POSITION
#800  EXTRACTOR FEED FAST RAMP FLOW LIMITS
#801  EXTRACTOR FEED FAST RAMP DELTA
#802  EXTRACTOR FEED SLOW RAMP FLOW LIMITS
#803  EXTRACTOR FEED SLOW RAMP DELTA
#804  EXTRACTOR FEED SPARE
#805  EXTRACTOR FEED SPARE
#806  EXTRACTOR #1 LEVEL MODE
#807  EXTRACTOR #1 LEVEL FLAG
#808  EXTRACTOR #1 LEVEL TARGET VALVE POSITION
#809  EXTRACTOR #1 LEVEL TARGET LEVEL
#810  EXTRACTOR #1 LEVEL COUNT SINCE LAST ADJUSTMENT
#811  EXTRACTOR #1 LEVEL LEVEL AT LAST ADJUSTMENT
#812  EXTRACTOR #1 LEVEL FUDGE FACTOR DELTA LEVEL TO DELTA VALVE POSITION
#813  EXTRACTOR #1 LEVEL CVP TO CVP MULTIPLIER
#814  EXTRACTOR #2 LEVEL MODE
#815  EXTRACTOR #2 LEVEL FLAG
#816  EXTRACTOR #2 LEVEL TARGET VALVE POSITION
#817  EXTRACTOR #2 LEVEL TARGET LEVEL
#818  EXTRACTOR #2 LEVEL COUNT SINCE LAST ADJUSTMENT
```

TABLE 6-4 continued

TABLE 6-4 continued

```
#819 EXTRACTOR #2 LEVEL LEVEL AT LAST ADJUSTMENT
#820 EXTRACTOR #2 LEVEL FUDGE FACTOR DELTA LEVEL TO DELTA VALVE POSITION
#821 EXTRACTOR #2 LEVEL CVP TO CVP MULTIPLIER
#822 EXTRACTOR #3 LEVEL MODE
#823 EXTRACTOR #3 LEVEL FLAG
#824 EXTRACTOR #3 LEVEL TARGET VALVE POSITION
#825 EXTRACTOR #3 LEVEL TARGET LEVEL
#826 EXTRACTOR #3 LEVEL COUNT SINCE LAST ADJUSTMENT
#827 EXTRACTOR #3 LEVEL LEVEL AT LAST ADJUSTMENT
#828 EXTRACTOR #3 LEVEL FUDGE FACTOR DELTA LEVEL TO DELTA VALVE POSITION
#829 EXTRACTOR #3 LEVEL CVP TO CVP MULTIPLIER
#830 FROG'S SPIT SOLVENT FEED MODE
#831 FROG'S SPIT SOLVENT FEED FLAG
#832 FROG'S SPIT SOLVENT FEED MASS FLOW TARGET
#833 FROG'S SPIT SOLVENT FEED RATIO HISTORY
#834 FROG'S SPIT SOLVENT FEED TARGET RATIO
#835 FROG'S SPIT SOLVENT FEED CURRENT RATIO
#836 FROG'S SPIT SOLVENT FEED CVP AT FIRST FLOW (VALVE OFFSET)
#837 FROG'S SPIT SOLVENT FEED FLOW AT ZERO CVP (FLOW OFFSET)
#838 FROG'S SPIT SOLVENT FEED TARGET VALVE POSITION
#839 FROG'S SPIT SOLVENT FEED RAMP DELTA FOR ADJUSTED RATIO
#840 FROG'S SPIT SOLVENT FEED MAXIMUM SINGLE CYCLE MASS FLOW ADJUSTMENT
#841 FROG'S SPIT SOLVENT FEED SPARE
#842 EVAPORATOR LEVEL MODE
```

TABLE 6-4 *continued*

```
#843 EVAPORATOR LEVEL FLAG
#844 EVAPORATOR LEVEL MASS FLOW TARGET
#845 EVAPORATOR LEVEL RATIO HISTORY
#846 EVAPORATOR LEVEL CURRENT RATIO
#847 EVAPORATOR LEVEL CVP AT FIRST FLOW (VALVE OFFSET)
#848 EVAPORATOR LEVEL FLOW AT ZERO CVP (FLOW OFFSET)
#849 EVAPORATOR LEVEL TARGET VALVE POSITION
#850 EVAPORATOR LEVEL HISTORY
#851 EVAPORATOR LEVEL MAXIMUM SINGLE CYCLE MASS FLOW ADJUSTMENT
#852 EVAPORATOR LEVEL LOW LIMIT
#853 EVAPORATOR LEVEL HIGH LIMIT
#854 EVAPORATOR LEVEL: MULTIPLIER, LEVEL DEVIATION TO MASS FLOW ADJUSTMENT
#855 EVAPORATOR LEVEL CVP TO CVP MULTIPLIER
#856 EVAPORATOR LEVEL SPARE
#857 EVAPORATOR TEMPERATURE MODE
#858 EVAPORATOR TEMPERATURE FLAG
#859 EVAPORATOR TEMPERATURE MASS FLOW TARGET
#860 EVAPORATOR TEMPERATURE RATIO HISTORY
#861 EVAPORATOR TEMPERATURE CURRENT RATIO
#862 EVAPORATOR TEMPERATURE CVP AT FIRST FLOW (VALVE OFFSET)
#863 EVAPORATOR TEMPERATURE FLOW AT ZERO CVP (FLOW OFFSET)
#864 EVAPORATOR TEMPERATURE TARGET VALVE POSITION
#865 EVAPORATOR TEMPERATURE MAXIMUM SINGLE CYCLE MASS FLOW ADJUSTMENT
```

TABLE 6-4 *continued*

TABLE 6-4 continued

TABLE 6-4 continued

```
#866 EVAPORATOR TEMPERATURE LOW LIMIT
#867 EVAPORATOR TEMPERATURE HIGH LIMIT
#868 EVAPORATOR TEMPERATURE OUT OF LIMIT SUCCESSIVE COUNTER
#869 EVAPORATOR TEMPERATURE OUT OF LIMIT MAX COUNT
#870 EVAPORATOR TEMPERATURE DELAY COUNTER
#871 EVAPORATOR TEMPERATURE DELAY COUNT
#872 EVAPORATOR TEMPERATURE: MULTIPLIER, TEMPERATURE DEVIATION TO MASS FLOW ADJUSTMENT
#873 EVAPORATOR TEMPERATURE SPARE
#874 TSJ FEED MODE
#875 TSJ FEED FLAG
#876 TSJ FEED MASS FLOW TARGET
#877 TSJ FEED RATIO HISTORY
#878 TSJ FEED TARGET RATIO
#879 TSJ FEED CURRENT RATIO
#880 TSJ FEED CVP AT FIRST FLOW (VALVE OFFSET)
#881 TSJ FEED FLOW AT ZERO CVP (FLOW OFFSET)
#882 TSJ FEED TARGET VALVE POSITION
#883 TSJ FEED RAMP DELTA FOR ADJUSTED RATIO
#884 TSJ FEED MAXIMUM SINGLE CYCLE MASS FLOW ADJUSTMENT
#885 DISTILLATION COLUMN FEED MODE
#886 DISTILLATION COLUMN FEED FLAG
#887 DISTILLATION COLUMN FEED LONG TERM MASS FLOW TARGET
#888 DISTILLATION COLUMN FEED SHORT TERM MASS FLOW TARGET
#889 DISTILLATION COLUMN FEED TARGET HISTORY
```

TABLE 6-4 continued

TABLE 6-4 continued

```
#890 DISTILLATION COLUMN FEED CVP AT FIRST FLOW (VALVE OFFSET)
#891 DISTILLATION COLUMN FEED FLOW AT ZERO CVP (FLOW OFFSET)
#892 DISTILLATION COLUMN FEED TARGET VALVE POSITION
#893 DISTILLATION COLUMN FEED FAST RAMP FLOW LIMITS
#894 DISTILLATION COLUMN FEED FAST RAMP DELTA
#895 DISTILLATION COLUMN FEED SLOW RAMP FLOW LIMITS
#896 DISTILLATION COLUMN FEED SLOW RAMP DELTA
#897 DISTILLATION COLUMN FEED SPARE
#898 DISTILLATION COLUMN FEED SPARE
#899 DISTILLATION COLUMN LEVEL MODE
#900 DISTILLATION COLUMN LEVEL FLAG
#901 DISTILLATION COLUMN LEVEL MASS FLOW TARGET
#902 DISTILLATION COLUMN LEVEL RATIO HISTORY
#903 DISTILLATION COLUMN LEVEL CURRENT RATIO
#904 DISTILLATION COLUMN LEVEL CVP AT FIRST FLOW (VALVE OFFSET)
#905 DISTILLATION COLUMN LEVEL FLOW AT ZERO CVP (FLOW OFFSET)
#906 DISTILLATION COLUMN LEVEL TARGET VALVE POSITION
#907 DISTILLATION COLUMN LEVEL MAXIMUM SINGLE CYCLE MASS FLOW ADJUSTMENT
#908 DISTILLATION COLUMN LEVEL LOW LIMIT
#909 DISTILLATION COLUMN LEVEL HIGH LIMIT
#910 DISTILLATION COLUMN LEVEL: MULTIPLIER, LEVEL DEVIATION TO MASS FLOW ADJUSTMENT
#911 DISTILLATION COLUMN LEVEL CVP TO CVP MULTIPLIER
#912 DISTILLATION COLUMN LEVEL SPARE
```

TABLE 6-4 continued

TABLE 6-4 continued

```
#913 DISTILLATION COLUMN REFLUX MODE
#914 DISTILLATION COLUMN REFLUX FLAG
#915 DISTILLATION COLUMN REFLUX MASS FLOW TARGET
#916 DISTILLATION COLUMN REFLUX RATIO HISTORY
#917 DISTILLATION COLUMN REFLUX TARGET RATIO
#918 DISTILLATION COLUMN REFLUX CURRENT RATIO
#919 DISTILLATION COLUMN REFLUX CVP AT FIRST FLOW (VALVE OFFSET)
#920 DISTILLATION COLUMN REFLUX FLOW AT ZERO CVP (FLOW OFFSET)
#921 DISTILLATION COLUMN REFLUX TARGET VALVE POSITION
#922 DISTILLATION COLUMN REFLUX MINIMUM FLOW
#923 DISTILLATION COLUMN REFLUX RAMP DELTA FOR ADJUSTED RATIO
#924 DISTILLATION COLUMN REFLUX MAXIMUM SINGLE CYCLE MASS FLOW ADJUSTMENT
#925 DISTILLATION COLUMN REFLUX SPARE
#926 DISTILLATION COLUMN STEAM MODE
#927 DISTILLATION COLUMN STEAM FLAG
#928 DISTILLATION COLUMN STEAM MASS FLOW TARGET
#929 DISTILLATION COLUMN STEAM RATIO HISTORY
#930 DISTILLATION COLUMN STEAM CURRENT RATIO
#931 DISTILLATION COLUMN STEAM CVP AT FIRST FLOW (VALVE OFFSET)
#932 DISTILLATION COLUMN STEAM FLOW AT ZERO CVP (FLOW OFFSET)
#933 DISTILLATION COLUMN STEAM TARGET VALVE POSITION
#934 DISTILLATION COLUMN STEAM MAXIMUM SINGLE CYCLE MASS FLOW ADJUSTMENT
#935 DISTILLATION COLUMN BOTTOM'S TEMPERATURE LOW LIMIT
#936 DISTILLATION COLUMN BOTTOM'S TEMPERATURE HIGH LIMIT
```

TABLE 6-4 continued

```
#937 DISTILLATION COLUMN BOTTOM'S TEMPERATURE OUT OF LIMIT SUCCESSIVE COUNTER
#938 DISTILLATION COLUMN BOTTOM'S TEMPERATURE OUT OF LIMIT MAX COUNT
#939 DISTILLATION COLUMN STEAM DELAY COUNTER
#940 DISTILLATION COLUMN STEAM DELAY COUNT
#941 DISTILLATION COLUMN TEMPERATURE: MULTIPLIER, TEMPERATURE DEVIATION TO STEAM FLOW ADJUSTMENT
#942 DISTILLATION COLUMN STEAM SPARE
#943 DISTILLATION COLUMN CONDENSER TEMPERATURE MODE
#944 DISTILLATION COLUMN CONDENSER TEMPERATURE FLAG
#945 DISTILLATION COLUMN CONDENSER TEMPERATURE MASS FLOW TARGET
#946 DISTILLATION COLUMN CONDENSER TEMPERATURE RATIO HISTORY
#947 DISTILLATION COLUMN CONDENSER TEMPERATURE CURRENT RATIO
#948 DISTILLATION COLUMN CONDENSER TEMPERATURE CVP AT FIRST FLOW (VALVE OFFSET)
#949 DISTILLATION COLUMN CONDENSER TEMPERATURE FLOW AT ZERO CVP (FLOW OFFSET)
#950 DISTILLATION COLUMN CONDENSER TEMPERATURE TARGET VALVE POSITION
#951 DISTILLATION COLUMN CONDENSER TEMPERATURE MAXIMUM SINGLE CYCLE MASS FLOW ADJUSTMENT
#952 DISTILLATION COLUMN CONDENSER TEMPERATURE LOW LIMIT
#953 DISTILLATION COLUMN CONDENSER TEMPERATURE HIGH LIMIT
#954 DISTILLATION COLUMN CONDENSER TEMPERATURE OUT OF LIMIT SUCCESSIVE COUNTER
#955 DISTILLATION COLUMN CONDENSER TEMPERATURE OUT OF LIMIT MAX COUNT
#956 DISTILLATION COLUMN CONDENSER TEMPERATURE DELAY COUNTER
#957 DISTILLATION COLUMN CONDENSER TEMPERATURE DELAY COUNT
#958 DISTILLATION COLUMN CONDENSER TEMPERATURE: MULTIPLIER, TEMPERATURE DEVIATION TO MASS FLOW ADJUSTMENT
#959 DISTILLATION COLUMN CONDENSER TEMPERATURE SPARE
```

TABLE 6-4 continued

TABLE 6-4 continued

TABLE 6-4 continued

#960	DOUBLE FEED SET-POINT POSITION ADJUSTMENT
#961	BUBBLE FEED SET-POINT POSITION ADJUSTMENT
#962	TOIL & TROUBLE FEED SET-POINT POSITION ADJUSTMENT
#963	REACTOR RECYCLE FLOW SET-POINT POSITION ADJUSTMENT
#964	REACTOR LEVEL SET-POINT POSITION ADJUSTMENT
#965	REACTOR TEMPERATURE SET-POINT POSITION ADJUSTMENT
#966	SOAKER PRESSURE SET-POINT POSITION ADJUSTMENT
#967	FLASH DRUM PRESSURE SET-POINT POSITION ADJUSTMENT
#968	FLASH DRUM CONDENSER TEMPERATURE SET-POINT POSITION ADJUSTMENT
#969	FLASH DRUM COOLER TEMPERATURE SET-POINT POSITION ADJUSTMENT
#970	FLASH DRUM LEVEL SET-POINT POSITION ADJUSTMENT
#971	NEUTRALIZER PH SET-POINT POSITION ADJUSTMENT
#972	NEUTRALIZER LEVEL SET-POINT POSITION ADJUSTMENT
#973	FIRST EXTRACTOR DRUM LEVEL SET-POINT POSITION ADJUSTMENT
#974	SECOND EXTRACTOR DRUM LEVEL SET-POINT POSITION ADJUSTMENT
#975	THIRD EXTRACTOR DRUM LEVEL SET-POINT POSITION ADJUSTMENT
#976	EXTRACTOR PRESSURE SET-POINT POSITION ADJUSTMENT
#977	SOLVENT FEED FLOW SET-POINT POSITION ADJUSTMENT
#978	EVAPORATOR LEVEL SET-POINT POSITION ADJUSTMENT
#979	EVAPORATOR TEMPERATURE SET-POINT POSITION ADJUSTMENT
#980	EVAPORATOR CONDENSER TEMPERATURE SET-POINT POSITION ADJUSTMENT
#981	EVAPORATOR CONDENSER LEVEL SET-POINT POSITION ADJUSTMENT
#982	TSJ TINCTURE FEED FLOW SET-POINT POSITION ADJUSTMENT
#983	DISTILLATION COLUMN FEED FLOW SET-POINT POSITION ADJUSTMENT
#984	DISTILLATION COLUMN LEVEL SET-POINT POSITION ADJUSTMENT

TABLE 6-4 continued

TABLE 6-4 continued

#985	DISTILLATION COLUMN REFLUX DRUM LEVEL SET-POINT POSITION ADJUSTMENT
#986	DISTILLATION COLUMN REFLUX FLOW SET-POINT POSITION ADJUSTMENT
#987	DISTILLATION COLUMN CONDENSER TEMPERATURE SET-POINT POSITION ADJUSTMENT
#988	DISTILLATION COLUMN BOTTOMS TEMPERATURE SET-POINT POSITION ADJUSTMENT
#989	SPARE
#990	SPARE
#991	SPARE
#992	DOUBLE FEED CONTROL-VALVE POSITION ADJUSTMENT
#993	BUBBLE FEED CONTROL-VALVE POSITION ADJUSTMENT
#994	TOIL & TROUBLE FEED CONTROL-VALVE POSITION ADJUSTMENT
#995	REACTOR RECYCLE CONTROL-VALVE POSITION ADJUSTMENT
#996	REACTOR FORWARD FLOW CONTROL-VALVE POSITION ADJUSTMENT
#997	REACTOR HEATER STEAM FLOW CONTROL-VALVE POSITION ADJUSTMENT
#998	SOAKER FORWARD FLOW CONTROL-VALVE POSITION ADJUSTMENT
#999	FLASH DRUM VENT CONTROL-VALVE POSITION ADJUSTMENT
#1000	FLASH DRUM CONDENSER WATER CONTROL-VALVE POSITION ADJUSTMENT
#1001	FLASH DRUM COOLER WATER CONTROL-VALVE POSITION ADJUSTMENT
#1002	FLASH DRUM FORWARD FLOW CONTROL-VALVE POSITION ADJUSTMENT
#1003	MARES' SWEAT FEED CONTROL-VALVE POSITION ADJUSTMENT
#1004	EXTRACTORS FEED CONTROL-VALVE POSITION ADJUSTMENT
#1005	FIRST EXTRACTORS DRUM CONTROL-VALVE POSITION ADJUSTMENT
#1006	SECOND EXTRACTORS DRUM CONTROL-VALVE POSITION ADJUSTMENT
#1007	THIRD EXTRACTORS DRUM CONTROL-VALVE POSITION ADJUSTMENT

TABLE 6-4 continued

#	
#1008	EXTRACT CONTROL-VALVE POSITION ADJUSTMENT
#1009	SOLVENT CONTROL-VALVE POSITION ADJUSTMENT
#1010	EVAPORATOR BOTTOMS CONTROL-VALVE POSITION ADJUSTMENT
#1011	EVAPORATOR REBOILER STEAM CONTROL-VALVE POSITION ADJUSTMENT
#1012	EVAPORATOR CONDENSER WATER CONTROL-VALVE POSITION ADJUSTMENT
#1013	EVAPORATOR FORWARD FLOW CONTROL-VALVE POSITION ADJUSTMENT
#1014	TSJ TINCTURE FEED CONTROL-VALVE POSITION ADJUSTMENT
#1015	DISTILLATION COLUMN FEED CONTROL-VALVE POSITION ADJUSTMENT
#1016	DISTILLATION COLUMN BOTTOMS FLOW CONTROL-VALVE POSITION ADJUSTMENT
#1017	DISTILLATION COLUMN FORWARD FLOW CONTROL-VALVE POSITION ADJUSTMENT
#1018	DISTILLATION COLUMN REFLUX CONTROL-VALVE-POSITION ADJUSTMENT
#1019	DISTILLATION COLUMN CONDENSER WATER CONTROL-VALVE POSITION ADJUSTMENT
#1020	DISTILLATION COLUMN REBOILER STEAM CONTROL-VALVE POSITION ADJUSTMENT
#1021	SPARE
#1022	SPARE
#1023	SPARE

Table 6-5
Required CRT Stations for Witches' Brew Plant

#1 Plant Manager
#2 Clerk
#3 Engineer
#4 Maintenance shop
#5 Control room alarms
#6 Control room control reaction
#7 Control room control finishing
#8 Control room graphic trending
#9 Control room various displays
#10 Control room various displays

Projected Costs

A. Computer Equipment—Process-Dedicated

Tweedledum (the process computer system)

1a. Two cabinets will be required in which to mount all the equipment, to include doors front and back, casters, and cooling fans ($1000 each) $2000

1b. One CPU including chassis, backplane, bus extensions, programmer's console, ALU processor, interface for console device, memory control, and power supplies $11,000

1c. The full complement of memory for the CPU $4000

1d. One floating-point processor $3000

1e. One option card to include power-fail auto-restart, a clock, and various bootstrap loaders $1000

1f. One hard-copy matrix teletypewriter as console device ... $2000

1g. One high-speed paper-tape reader punch with interface .. $3500

1h. One twin-drive double-density floppy-disk system and interface $3500

2a. One very large mass storage moving-head disk system of 40 megabytes $9000

2b. One controller and interface for disk $4000

3a. One serial communications interface for 16 multiplexed channels.. $12,000

Table 6-6
Proposed Computer Control Loops

1. Double feed flow
2. Bubble feed flow
3. Toil and Trouble feed flow
4. Reactor recycle flow
5. Reactor level
6. Reactor temperature
7. FD Condenser temperature
8. FD Cooler temperature
9. FD level
10. Neutralizer pH
11. Neutralizer level
12. Extractor #1 level
13. Extractor #2 level
14. Extractor #3 level
15. Solvent feed flow
16. Evaporator level
17. Evaporator temperature
18. TSJ flow
19. Distillation column feed
20. Distillation column level
21. Distillation column reflux
22. Distillation column temperature
23. Distillation column condenser temperature

3b. One 19-inch, 8-color CRT, 19.2K Baud RS232 capability, with 80-column × 48-row character display, supporting 128 ASCII character set, and including keyboard with 16 special function keys ($3000/unit) $3000

3c. Nine 12-inch, b&w CRTs, 19.2K Baud RS232 capability, with 80-column × 24-row character display, supporting 128 ASCII character set, and including keyboard with numeric pad, character accents of blinking, dim, and reverse video ($1000 each) $9000

3d. One high-speed (300 lpm) 132-column matrix printer/ plotter with parallel capability. ($6000 each) $6000

3e. One parallel interface and cables for printer/plotter ($1000 each) $1000

4a. One universal "front-end" controller and interface for analog input channel selection, gain setting, and multiplexing (but no individual input channels) and digital input digital output channel selection with bit setting or bit reading of DI/DO words (but no individual input output channels) ($3000) $3000

4b. 320 analog input channels as 40 12-bit resolution, 8-channel, signal conditioning signal converting analog input cards. ($50/channel; $400/each card) $16,000

4c. 160 DIDO signals as 10 16-bit signal conditioning digital input and digital output cards ($10/bit; $160/card) $1600

B. Computer Equipment—Non-Process-Dedicated

Tweedledee (the development and testing computer system)

1a. Two cabinets will be required in which to mount all the equipment, including doors front and back, casters, and cooling fans. ($1000 each) $2000

1b. One CPU including chassis, backplane, bus extensions, programmer's console, ALU processor, interface for console device, memory control, and power supplies. ($11,000 total) $11,000

1c. The full complement of memory for the CPU $4000

1d. One floating-point processor $3000

1e. One option card to include power-fail auto-restart, a clock, and various bootstrap loaders $1000

1f. One hard-copy matrix teletypewriter for console device .. $2000

1g. One high-speed paper-tape reader punch with interface .. $3500

1h. One twin-drive double-density floppy-disk system and interface $3500

2a. One very large mass storage moving-head disk system of 40 megabytes $9000

2b. One controller and interface for disk $4000

3a. One serial communications interface for 16 multiplexed channels. ($12,000 per unit) $12,000

3b. One 19-inch, 8-color CRT, 19.2K Baud RS232 capability, with 80-column × 48-row character display, supporting 128 ASCII character set, and including keyboard with 16 special function keys. ($3000/unit) $3000

3c. One 12-inch, b&w CRT, 19.2K Baud RS232 capability, with 80-column × 24-row character display, supporting 128 ASCII character set and including keyboard with

numeric pad, video character accents of blinking, dim, and
reverse video. ($1000/unit) $1000

3d. The high-speed (300 lpm) 132-column matrix printer/
plotter on Tweedledum will be shared between the two
systems on an "as needed" basis

3e. One parallel interface and cables for printer/plotter $1000

4a. One universal "front-end" controller and interface for
analog input channel selection, gain setting, and
multiplexing (but no individual input channels), and
digital-input digital-output channel selection with bit-
setting or bit-reading of DI/DO words (but no individual
input output channels) $3000

4b. Eight analog input channels as one 12-bit resolution 8-
channel signal conditioning signal converting analog input
card. ($50/channel; $400/card) $400

4c. 32 DIDO signals as two 16-bit signal conditioning digital-
input digital-output cards. ($10/bit; $160/card) $320

4d. 40 pairs of screw terminals plus necessary cabling
($20/pair) $800

C. Software, Fees, and Training (costs and summary)

1. Consultants' fees to spec, purchase, take delivery, install,
and check out @ 2.5% of gross hardware cost $3978

2. Manufacturers' software for Tweedledee (includes listings,
hot-line support, and updates for one year) $6000

3. Two man-months of factory training on hardware and manufacturers' software (includes airfare, per diem, hotels,
rental car, and fees) $15,000

4. Three man-months (trained person) customizing system
software from previously proven operating system for the
process computer (Tweedledee) $10,500

5. Six man-months (trained person) implementation,
debugging, and proving of customized application
software checking out all analog inputs, digital inputs,
digital outputs $21,000

6. Six man-months (trained person) to close 24 loops. One
man-week per loop on average $21,000

Total $77,478

Note: Six man-months are required to train a sufficiently motivated engineer to "work alone" competency. Twelve man-months are required to write a completely new operating system from scratch by a trained person. Salary and benefits for chemical engineers with minicomputer experience, office space, and overheads: $3500/man-month.

D. Instrumentation Modifications Per Item Costs

1. Analog Inputs

Process variables must be presented to the computer's front-end as electric analogs. This will require installing transmitters for those variables that are to be connected to the front-end and currently do not have transmitters. Any pneumatic transmitters that are already installed or are to be installed will require a pneumatic-to-voltage transducer. Electronic transmitters will not require this supplementary device.

The following costs are an approximate total for such equipment, labor, miscellaneous small parts, and cabling required to provide one electric analog at the computer installation's field-wiring termination panel (FWTP) for the various listed alternatives.

1a. Existent pneumatic transmitter. A pneumatic-to-voltage
transducer, labor, cabling $500

1b. Existent electronic transmitter (includes thermocouples).
Labor, cabling $200

1c. New pneumatic transmitter. Pneumatic transmitter,
transducer, labor, cabling $1300

1d. New electronic transmitter. Electronic transmitter, labor,
cabling $800

2. Controllers

It is usual to have analog controllers between the computer and the process. They provide back-up control should the computer malfunction and have to be taken off-line. They also enable the operators to override the computer should they deem it necessary. There is no significant cost difference between SSC and DDC implementation. Consequently, only one cost will be given for each of the following alternatives. As before, the costs are an approximate total for such equipment, labor, miscellaneous small parts, and cabling as would be required for one controller and includes the termination of the remote-local status switch, pulse-up, pulse-down, computer-fail signals at the DI/DO terminals, and the set-point or valve-position analog signal at the AI terminal within the FWTP. These costs are exclusive of providing an analog measurement of the controlled variable to the FWTP.

2a. Existent computer-incompatible pneumatic controller with combined recorder. New computer compatible pneumatic controller that includes some form of pneumatic-to-voltage transducing arrangement for the computer signals to adjust the set-point/valve-position, new multipen recorder, labor, cabling $3500

2b. Existent computer-incompatible pneumatic controller without combined recorder. New computer-compatible pneumatic controller that includes some form of pneumatic-to-voltage transducing arrangement for the computer signals to adjust the set-point/valve-position, labor, cabling $2500

2c. Existent computer-compatible pneumatic controller (transduced set-point signal available). Labor, cabling ... $400

2d. Existent pneumatic controller that can take computer-compatible conversion kit. Conversion kit, labor, cabling $1000

2e. Existent computer-incompatible electronic controller. New computer-compatible electronic controller, labor, cabling $2400

2f. Existent computer-compatible electronic controller. Labor, cabling $400

2g. Existent electronic controller that can take computer-compatible conversion kit. Conversion kit, labor, cabling $800

2h. Replacement of existent pneumatic controller with computer-compatible electronic controller, but retaining pneumatic field transmitter and pneumatic recorders. New pneumatic-to-current transducer (incoming signal from field transmitter), new computer-compatible electron controller, new current-to-pneumatic transducer (outgoing signal to control valve), labor, cabling $2800

2i. Replacement of existent pneumatic controller and pneumatic transmitter with computer-compatible electronic controller and electronic transmitter, but retaining pneumatic recorder. New electronic transmitter, new computer-compatible electronic controller, new

current-to-pneumatic transducer (outgoing signal to
control valve), new current-to-pneumatic transducer
(incoming signal to recorder), labor, cabling $3500

2j. Replacement of existent pneumatic equipment with
computer-compatible electronic controller, new electronic
transmitter, and electronic recorder. One electronic
transmitter, one electronic controller, one electronic
recorder, one current-to-pneumatic transducer
(outgoing signal to control valve), labor, cabling $4800

3. Miscellania

If it were found necessary to install a new thermowell or new orifice flanges
and plate for a metering station, such costs would be in addition to those
outlined previously. Also, the previous costs do *not* include new instrument
shelving or new panel boards. If they are necessary, that will be at additional
cost.

Total Projected Costs

These are summarized below:

Computer #1 (Tweedledum) .... $94,600   (Table 6-7)
Computer #2 (Tweedledee) 64,520   (Table 6-8)
Software, Fees, Training 77,478   (Section C)
Instrumentation 160,500   (Table 6-9)

Grand Total $397,098

Notes:

1. One CPU alone without any internal or external options costs $11,000, or about
    3% of overall project cost, yet the total computer hardware is 40% of project.

2. Software costs for the project are about equal to the cost of a system configured
    for process control, including all peripherals. In this example software is about
    20% of total project.

3. Instrument modification is about 40% of the total project.

4. In very round numbers a single user non-process control system
    can be configured for $20,000

Very large mass storage added.. 15,000

Enhancements to improve performance............................... 5,000

Multichannel human interface plus peripherals 20,000

Chemical process-oriented front-end 20,000

$80,000

Table 6-7
Tweedledum (Process-Dedicated Total Cost Summary)

Items	Cost($)	Subtotals
Cabinetry (2)	2,000	
CPU (Power supplies)	11,000	
Memory	4,000	
FPP	3,000	
PFAR, Clock, Bootstraps	1,000	
Teletypewriter console	2,000	
Paper-tape reader/punch	3,500	
Floppy-disk drives	3,500	
		30,000
Very large mass storage disk system	9,000	
Disk controller	4,000	
		13,000
Communications I-face 16ch MUX	12,000	
Color CRTS (@ $3000 each)	3,000	
B&W CRTS (9 @ $1000 each)	9,000	
		24,000
High-speed line printer/plotter	6,000	
Interface for printer/plotter	1,000	
		7,000
Front-end controller	3,000	
Analog inputs (320 @ $50 each)	16,000	
Digin/digout (160 @ $10 each)	1,600	
		20,600
		Total: $94,600

The two actual bare-bones single-user computer systems in this project are about 10% of the total project. Upgrading both to powerful machines is another 10% of the project. Making them multiuser by adding extra human interfaces is a further 10% of the project. The addition of a chemical process-oriented front-end in both cases adds another 10% to the project.

Projected Savings

As mentioned before, the savings will fall into one of two classes: those savings that accrue while the computer takes a passive role (that is, before the first computerized control loop is closed), and those savings that are generated when the computer takes an active role (that is, the savings created when a computer actively supervises and controls the process).

Table 6-8
Tweedledee (Non-Process-Dedicated Total Cost Summary)

Items	Cost($)	Subtotals
Cabinetry (2)	2000	
CPU (Power supplies)	11,000	
Memory	4000	
FPP	3000	
PFAR, Clock, Bootstraps	1000	
Teletypewriter console	2000	
Paper-tape reader/punch	3500	
Floppy-disk drive	3500	
		30,000
Very large mass storage disk	9000	
Disk controller	4000	
		13,000
Communications I-face 16ch MUX	12,000	
Color CRT (1 at $3000)	3000	
B&W CRT (1 @ $1000)	1000	
		16,000
Interface for printer/plotter	1000	
		1000
Front-end controller	3000	
Analog inputs (8 @ $50 each)	400	
Digin/digout (32 @ $10 each)	320	
Screw terminals & cabling (40 @ $20 each)	800	
		4520
		Total: $64,520

The passive-role savings stem from the discipline of calibrating all transmitters for correct zero and correct span; the double-checking of all orifice plate factors; the overall greater involvement of engineers and operators with the process; and finally, the most important contribution—the consideration of certain derived control parameters such as reflux ratio, lb/day of product in waste streams, and reactor stoichiometry that are now calculated effortlessly and are instantaneously accessible. Additional savings come from less conservative brinkmanship as knowledge of the process grows and those minor physical and mental bottlenecks pinpointed by using the computer as a process investigation tool are removed.

Table 6-9
Instrumentation Modifications (Total Costs and Summary)

Items	Cost($)
31 analog inputs (pneumatic transmitters existent)	$15,500
44 analog inputs (new electronic transmitters)	35,200
62 analog inputs (electronic transmitters existent)	12,400
12 control loops (replacing incompatible pneumatic controllers)	30,000
11 control loops (replacing pneumatic loop with all electronic equipment)	52,800
1 field-wiring termination panel (FWTP), terminals, and cabling to processor (480 signals at $20 each)	9,600
1 control room console	5,000
	Total: $160,500

The active-role savings are in part straightforward raw material savings as better reactor stoichiometry increases the yields. They are in part reduced utility consumption as less material is off spec, requiring less material to be rerun; they are in part plant productivity as the capacity of the plant has been increased. Side benefits of implemented computer control are an increase in the knowledge of the physical plant and the chemical process. There is better overall instrumentation, as any malfunctioning control valves or unsatisfactory analog controllers are diagnosed easily, and control is perhaps orchestrated around the problem until it is solved. Also, previously unrecognized event interaction between flows and/or equipment may be discovered. The beneficial interaction can be promoted and the degrading interaction suppressed. With the essentially feed-forward nature of the computer control, fewer disturbances actually manifest themselves within the system. Those that do occur are dampened more quickly, and their side effects may be expected and better compensated. Overall the process will be less prone to erratic behavior, and it will become a smoother operation.

The power of the computer to calculate allows the plant to be run at either an economic cost minimum or an economic profit maximum, which directly brings about a dollar benefit. With computer control, the brinkmanship becomes dynamic. The loading on any equipment can be increased or decreased automatically. Process throughput can be reduced during periods of upset conditions. When the correct conditions are reestablished, the previous throughput can be resumed.

In a four-shift non-computer plant each lead operator will have his fancy as where best to run each set point within the plant. It will not be a process optimum, but a personal comfort optimum set in such a fashion that an excursion will not cross that line deemed by his shift foreman as unacceptable operation. Each shift foreman draws his personal line such that excursions beyond it do not cross the line emphasized by management as the limit of good operation. The result is a process continuously cycled through four control stratagems, three of which are inferior to the fourth, which itself may be set well away from the most suitable and practicable economic criteria.

For the Witches' Brew Process, the possible savings for the two stages are estimated to be as follows:

Passive class

1. A 20% reduction in yield losses
2. A 5% reduction in utility consumption
3. A 10% increase in the plant capacity

Active class

1. A further 20% reduction in yield losses
2. A further 5% reduction in utility consumption
3. A further 10% increase in the plant capacity

The original base plant process economics are presented in Figure 6-1.

For the passive class of savings, the 20% reduction in yield losses is taken as a straightforward raw material yield increase on Double from 80% to 84%, on Bubble from 85% to 88% and on Toil and Trouble from 90% to 92%. The use of the raw materials is reduced correspondingly. Mares' Sweat consumption is reduced considerably as the by-product acid production is cut 20%. The 5% reduction in utility consumption is taken on steam and water but not on electricity. The possible increase in capacity is not used. The base plant process economics altered to include the passive savings are presented in Figure 6-2. The material and cost savings are presented in Table 6-10.

For the active class of savings, an opportunity is taken to increase the overall production rate, as it is the Toil and Trouble delivery system that is the capacity-limiting bottleneck. Consequently, the 20% reduction in yield loss for Toil and Trouble, which improve Toil and Trouble yield from 92% to 93.6% and would have dropped Toil and Trouble's rate of consumption from 11,739Mlb/y to 11,538Mlb/y, is used to increase production 3.8% by increasing Toil and Trouble consumption to 11,977Mlb/y. For Bubble, the yield is improved to 90.4% and consumption increased to 29,280Mlb/y. For Double, the yield comes up to 87.2%, and consumption is maintained at 4761Mlb/y. Mares' Sweat consumption is again cut 20% as the losses to the by-product acid are reduced 20%. However, as the overall production rate has increased 3.8%, the actual consumption of Mares' Sweat is 1993Mlb/y.

```
(i) total overall margin $873M, ROIBT 21.6%

           ANNUAL COSTS                                  CAPITAL
VARIABLE:                                                             M$
Raw Materials   YLD   c/#  M#/y   M$/y   M$/y    direct fixed        3000
DOUBLE         80.0%   5   5000   250            utilities            474
bubble         85.0%   6  30000  1800            other,admin, & sales 100
t & t          90.0%   2  12000   240            working,half month RM 114
m'sweat                15   3000   450           working,one month prod 362
               Total Raw Materials      2740
Utilities:                                            TOTAL CAPITAL 4050
steam     3$/M#      102000 M#/y     306
c'water   1$/10Mg    470000 Mg/y      47            SALES & PROFITS
elec      12$/Mkwh     2000 Mkwh/y    24                    M#/y   c/#   M$/y
               Total Utilities        377     Witches Brew  24000    18  4320
Manpower   14 people @ 22000$/y       308     EPR           10000    10  1000
Supplies  various                     298
               TOTAL VARIABLE COSTS   3723            TOTAL SALES        5320
FIXED:                                                            M$/y
maintenance      @  4% of DFC    120           total sales     5320
depreciation     @ 10% of DFC    300           total costs     4345
taxes & ins      @  2% of TC      81           selling expense  102
admin & factory  @  3% of TC     121                  MARGIN     873
               TOTAL FIXED COSTS      622
                                              ROIBT 873/4050 = 21.6%
               TOTAL ANNUAL COSTS     4345
```

Figure 6-1. Base plant process economics.

After the reduction allowed by the improved utility yields and the increase because of raised production rates, the utility consumption drops a net 1.4%. Witches' Brew production improves from 24,000 to 24,900Mlb/y, and EPR increases from 10,000 to 10,400Mlb/y. The base plant process economics with the inclusion of both the active and the passive savings are presented in Figure 6-3. The results are summarized in Table 6-11.

Reexamination of Process Economics

For a second look at the Witches' Brew Process economics, the full capacity increase in the plant is used, an overall 21% capacity gain (Table 6-12). However, to debottleneck those critical areas as pinpointed by the computer-aided process research, $112,000 will have to be spent on pump changes, piping modifications, and vessel alterations. Both the plant debottlenecking costs and the computer-related expenditures are rolled into the direct fixed cost of the plant, raising it to $3509M. The expanded plant process economics, including both the active and passive savings, the full capacity gain, and the extra capital, are presented in Figure 6-4.

For a base plant costing $3MM, a second plant (expanded linearly in process equipment costs for a 21% larger capacity but without a computer) could have been expected to cost $3600M.

Table 6-10
Passive Role Savings for Witches' Brew Process

Annual Costs	Base Case			After Installation		
	YLD	Mlb/y	M$/y	YLD	Mlb/y	M$/y
Double	80.0%	5,000	250	84.0%	4,762	238
Bubble	85.0%	30,000	1,800	88.0%	28,977	1,738
T & T	90.0%	12,000	240	92.0%	11,739	235
Mares' sweat		3,000	450		2,400	360
Steam		102,000	306		96,900	291
C'water (Mg/y)		470,000	47		446,000	45
Elec(MkWh/y)		2,000	24		2,000	24
Manpower			308			308
Supplies			298			298
Total variable costs			$3,723			$3,537
Total fixed costs*			622			620
Total annual costs			4,345			4,157

Results over base case:

Annual cost savings of $188M

ROIBT *INCREMENTAL* investment 188/397 = 47.4%

*The allocated FIXED COSTS are reduced because of reduced working capital requirements.

```
(i)     no production increase
(ii)    the DFC is unchanged from base case
(iii)   20% reduction in yield losses
(iv)    5% reduction in steam and water usage
(v)     results : overall costs savings of 188M$/y

              ANNUAL COSTS                                  CAPITAL
VARIABLE:                                                                    M$
Raw Materials   YLD    c/#   M#/y   M$/y   M$/y   direct fixed          3000
DOUBLE          84.0%   5    4762    238           utilities             450
bubble          88.0%   6   28977   1738          other,admin, & sales  100
t & t           92.0%   2   11739    235          working,half month RM 107
m'sweat                 15   2400     360          working,one month prod 346
              Total Raw Materials         2571
Utilities:                                                TOTAL CAPITAL 4003
steam      3$/M#       96900 M#/y    291
c'water    1$/10Ms    446000 Ms/y     45                 SALES & PROFITS
elec       12$/Mkwh     2000 Mkwh/y   24                              M#/y   c/#   M$/y
              Total Utilities           360  Witches Brew   24000    18   4320
Manpower  14 people @ 22000$/y          308  EPR            10000    10   1000
Supplies  various                       298
              TOTAL VARIABLE COSTS     3537           TOTAL SALES        5320
FIXED:                                                               M$/y
maintenance     @  4% of DFC    120          total sales        5320
depreciation    @ 10% of DFC    300          total costs        4157
taxes & ins     @  2% of TC      80          selling expense     102
admin & factory @  3% of TC     120                 MARGIN       1061
              TOTAL FIXED COSTS         620
                                             ROIBT 1061/4003 = 26.5%
          TOTAL ANNUAL COSTS   4157
```

Figure 6-2. Base plant process economics with passive-role savings.

```
(i)     a 3.8% production increase
(ii)    the DFC is unchanged from base case
(iii)   a further 20% reduction in yield losses
(iv)    a further  5% reduction in utility usage before production increase
(v)     results : increased sales of 202M$/y plus cost savings of 42M$/y

              ANNUAL COSTS                                  CAPITAL
VARIABLE:                                                                    M$
Raw Materials   YLD    c/#   M#/y   M$/y   M$/y   direct fixed          3000
DOUBLE          87.2%   5    4761    238           utilities             444
bubble          90.4%   6   29280   1757          other,admin, & sales  100
t & t           93.6%   2   11977    240          working,half month RM  106
m'sweat                 15   1993     299          working,one month prod 343
              Total Raw Materials         2534
Utilities:                                                TOTAL CAPITAL 3993
steam      3$/M#       95553 M#/y    287
c'water    1$/10Ms    440000 Ms/y     44                 SALES & PROFITS
elec       12$/Mkwh     1972 Mkwh/y   24                              M#/y   c/#   M$/y
              Total Utilities           355  Witches Brew   24900    18   4482
Manpower  14 people @ 22000$/y          308  EPR            10400    10   1040
Supplies  various                       298
              TOTAL VARIABLE COSTS     3495           TOTAL SALES        5522
FIXED:                                                               M$/y
maintenance     @  4% of DFC    120          total sales        5522
depreciation    @ 10% of DFC    300          total costs        4115
taxes & ins     @  2% of TC      80          selling expense     106
admin & factory @  3% of TC     120                 MARGIN       1301
              TOTAL FIXED COSTS         620
                                             ROIBT 1301/3993 = 32.6%
          TOTAL ANNUAL COSTS   4115
```

Figure 6-3. Base plant process economics with passive-role and active-role savings.

Table 6-11
Active Role Savings for Witches' Brew Process (Summary)

Annual Costs	Base Case			Passive		Active (No Increase)		Active (3.8% Increase)		
	YLD	Mlbs/y	M$/y	YLD	Mlb/y	YLD	Mlb/y	YLD	Mlb/y	M$/y
Double	80.0%	5,000	250	84.0%	4,762	87.2%	4,587	87.2%	4,761	238
Bubble	85.0%	30,000	1,800	88.0%	28,977	90.4%	28,208	90.4%	29,280	1,757
Toil & Trouble	90.0%	12,000	240	92.0%	11,739	93.6%	11,539	93.6%	11,977	240
Mares' Sweat		3,000	450		2,400		1,920		1,993	299
Steam		102,000	306		96,900		92,100		95,553	287
C'water(Mg/y)		470,000	47		446,000		423,700		440,000	44
Elec(MkWh/y)		2,000	24		2,000		1,900		1,972	24
Manpower			308							308
Supplies			298							298
Total variable costs			3,723							3,495
Total fixed costs			622							620
Total annual costs			4,345							4,115
Witches' brew		24,000	4,320		24,000		24,000		24,900	4,482
EPR		10,000	1,000		10,000		10,000		10,400	1,040
Total annual sales			5,320							5,522

Results over base case:

Annual cost savings of $230,000
Annual increased sales of $202,000
ROIBT *INCREMENTAL* investment 432/397 = 108.8%

Results case by case	Base		Passive		Active (100%)		Active (103.8%)	
	Total	Change	Total	Change	Total	Change	Total	Change
Annual costs	4,345		4,157	−188	4,005	−152	4,115	+110
Annual sales	5,320		5,320	0	5,320	0	5,522	+202

Table 6-12
Base Plant and Expanded Plant Comparisons

		Base	Expanded
Production "Witches' Brew," MMlbs		24.0	29.0
EPR, MMlbs		10.0	12.1
Direct fixed capital, M$		3,000	3,509
Pro rata share utility capital, M$		474	518
Pro rata share other, admin & sales, M$		100	100
Total working capital, M$		476	514
	Total capital:	$4,050	4,641
Total sales, M$/y		5,320	6,430
Total costs, M$/y		4,345	4,695
Selling expense, M$/y		102	123
	Margin, M$/y	873	1,612
	ROIBT, %	21.6	34.7

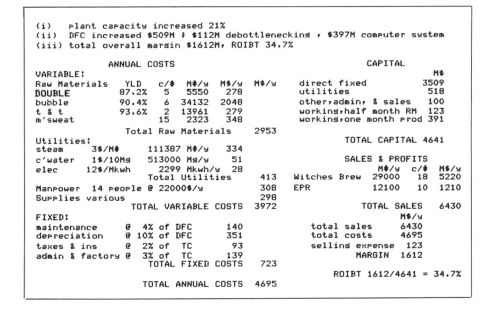

Figure 6-4. Expanded plant process economics.

This becomes an interesting proposition. In expanding the plant should the unfamiliar be done and the extra $509M over the base case be put into silicon-chip-related expenditures, or should the familiar be done and the extra $600M over the base case be spent on steel-fabric-related expenditures?

These exercises have been worked at length to emphasize several factors (see Table 6-13). First, payout time on computer-related projects is short. The longest payout period of the three scenarios is 25 months. Second, there is a multitude of ways that computer-induced savings can be calculated, but there is one saving often overlooked. Computers are cheap capacity. Finally, as computers can sometimes be bought off the shelf or even "previously owned" (whereas vessels, piping, and structure have a long gestation with shop fabrication and then field erection), the investment in a computer will find extra capacity very quickly for which, incidentally, no process equipment has to be cleared for the tie-ins to be made.

The estimates of the possible savings are optimistic approximations. A grossly inefficient plant will do much better than a plant already finely tuned. In either case the amount of possible savings is a function of the ingenuity and the persistence of the people involved. As emphasized elsewhere in this book, enthused and dedicated people will improve the system further and faster than will unwilling, coopted personnel. At least with a computer, documentable proof of improvements or lack of improvements is obtained easily.

If a tower is permitted to make product with specifications a generous margin away from sales specifications, then product is literally being given away. The overly generous portion of the prime component in the better product could be replaced by the lesser component, thereby upgrading that amount from the selling price of the inferior product to the selling price of the better product. In similar fashion the overly generous portions of prime component in the inferior product could be replaced by the lesser component, freeing more prime component for sale as the superior product, or generating the same sales quantities with less prime product. Large tonnage production plants could support a total computer system simply by reducing product giveaway. Table 6-14 is an example for the Witches' Brew.

So the economics say, "install a computer." The next task is to recruit a project manager. All too often the purchase of a computer system follows a very persuasive sales pitch either by a computer salesman pushing hardware, or a programmer who is pushing software. Part of the sales pitch will be directed toward emphasizing the attributes of computer control. Very little of the pitch will be directed to the attention to detail required and the difficulties that will be encountered. Assuming that the system is delivered, installed, genned, and the initial wrinkles debugged, the customer can expect the salesman or programmer to become involved with his next customer and he, the original customer, is on his own to develop further his process and

Table 6-13
Overall Costs and Savings with Payout

		M$	M$/y	Payout (months)
Investment:				
	Computer System	397		
Payback:				
Passive				
	Cost savings Passive		188	
	Increased sales		0	
	Total improvement to margin:	.	188	25
Active (no production increase)				
	Cost savings passive		188	
	Additional savings active		152	
	Total savings:		340	
	Increased sales		0	
	Total improvement to margin:		340	14
Active (production increase)				
	Cost savings passive		188	
	Additional savings active		152	
	Increased production costs		−110	
	Total net savings:		230	
	Increased sales		202	
	Total improvement to margin:		432	11

Table 6-14
Product Giveaway Penalties

	Nominal Production	Component	Actual production Mlb/yr		Sales Specifications Mlb/yr		Give Away
Witches' Brew	24,000	π	99%	23,760	98%	23,520	240,000lb of π
		ϕ	1%	240	2%	480	@8¢/lb
							=$19,200/yr
EPR	10,000	π	10%	1,000	5%	500	500,000lb of π
		ϕ	90%	9,000	95%	9,500	@8¢/lb
							=$40,000/yr
		Total π		24,760		24,020	
		total ϕ		9,240		9,980	

computer control. Therefore, it will be necessary for a plant manager or a process supervisor instituting the installation of a computer at his own behest to retain his very own chemical engineer/computer expert, perhaps in an architect-like capacity. If this is possible, then in the same way that an architect guarantees his client satisfaction and tailors available goods and services into a residence or office, so will the chemical engineer/computer expert guarantee a measure of performance and tailor available processors, peripherals, instruments, and software into the existing chemical process. The outcome of his efforts is a measurable economic improvement in the chemical's manufacture.

With or without his expert, the buyer must be aggressive, and he must ask searching questions. The ability to ask searching questions comes from having read widely on the subject of minicomputers and process control. The buyer should endeavor to get written answers to his questions and to get personal guarantees from software vendors. To the vendors it may be just one more system with which to be concerned—to the buyer it is the ONLY system with which to be concerned. Buyers should never accept at face value comprehensive blandishments about the product, such as "controls up to 128 loops, is capable of expansion to 320 analog inputs, provides for 255 user programs, is capable of supporting 16 users. . . ." He should go back and ask if any one system has ever been expanded to 128 loops. How many complete scans in a minute are possible when expanded to 320 inputs? How big can each individual program be? Can there be 16 users simultaneously? At 9600 Baud? He should arrange for a demonstration of software packages, sit down at a keyboard, and actually walk-through the use of the program. The buyer should ask for estimates of man-months that other customers have expended implementing satisfactory solutions with the vendor's products. Finally, best of all, the prospective buyer should make visits to other minicomputer process control installations and talk with users, both satisfied and disgruntled. He should endeavor to establish whether or not what he has seen will do what he wants done in his own plant. Only when he is well satisfied should he sit down and write purchase orders.

Presuming that the system is a profitable investment, the hardware and system software can be ordered. While waiting for delivery, continue with the pencil and paper.

Next, lay out each display in turn. This ensures that every quantity displayed will be either measured or calculated. It will be to the system's overall advantage if all these instantaneous variable values are held in common in memory. This will minimize data transfer and disk accesses. It also will ensure that the same one value of any variable is available to all programs in any one program cycle. A common of 1024 values will be created to hold all the AI inputs at their proper engineering values (#000-#319) and all the necessary factors and calculated variables (#320-#1023) (see Table 6-4).

All the calculations now can be laid out in pencil and paper in terms of the index numbers. The format strings for the individual displays can be generated. Each individual field in a display will need to have several features defined: its row and column position; whether the field is a non-numeric heading or a number; where the number or heading can be found in memory; how many columns wide the field is; how many decimal places the number should have; whether the number is to be represented as a decimal integer, an octal integer, or a real number. The conversion subroutines necessary to convert the digital values (DVALS) of the incoming analog measurements to their engineering values (EVALS) also can be listed (see Table 6-15).

When all these preliminary pencil and paper lists are complete and the computer systems delivered, the serious programming can begin. Complete the data gathering programs and data displaying programs first. Begin to look at the process the way the computer sees it. Develop control expertise carefully and slowly (see Chapter 8). How far and how fast the system and process are developed from this point is up to your imagination and initiative.

For Tweedledum, within the typical 10-second program cycle:

1. The 320 analog inputs are scanned, measured, and stored as digital values (DVALS).

2. The DVALS are converted to engineering values (EVALS).

3. The mass flows are calculated.

4. The control loops are checked individually. Those not on the computer are initialized with the current values of the relevant variables. This provides bumpless transfer when they are switched to the computer. Those loops on the computer have any necessary adjustments calculated and the necessary signals issued.

5. Each individual alarm point is checked. Any found to be out of the limits are flashed on the alarm screen. Those that have returned to within limits are erased from the screen.

6. All remaining variable calculations necessary for instantaneous values are completed.

7. All the CRT displays are refreshed with new information where applicable.

8. Any variables that are being plotted or tracked in batch mode have their current values packed into the requisite storage datasets.

9. Any elapsed-time counters that are being kept are updated as necessary.

10. The daily consumption calculations are updated for this 10-second period. *Text continued on page 186*

Table 6-15

Analog Input Subroutine Conversions for Tweedledum Software Package

| Subroutine | Engineering Values | | | Volts | | |
	Low	High		Low	High	
01	0	100	LINR	0	5	psig, degC, level
02	0	1000	LINR	1	5	Set point position, control valve
03	0	10	SQRT	0	5	Divisions flow pneumatic
04	0	10	SQRT	1	5	Divisions flow electronic
05	0	50	LINR	0	5	degC
06	0	8	LINR	0	5	Pressure psig
07	0	60	LINR	0	5	Pressure psig
08	7	14	LINR	1	5	pH
09	0	200	LINR	0	5	
10	40	60	LINR	0	5	Bubble purity
11	50	90	LINR	0	5	Toil & Trouble purity
12	0	1000	LINR	0	1	Corrosion probe
13	1.000	1.200	LINR	1	5	OP
14		Thermocouples				
15	0	300	LINR	0	5	
16	0	500	LINR	0	5	
17	0	5	LINR	1	5	
18	0	15	LINR	0	5	
19		High-temp thermocouples				
20	0	10	LINR	0	50mV	
21	0	.5	LINR	0	10mV	(0-.5%)
22	0	10	LINR	1	5	Vent %

23	0	15	LINR	1	5	inhg
24	0	500	LINR	1	5	Vent vppm
25	0	100	LINR	1	5	Level
26	0	800	LINR	0	5	psig
27	0	30	LINR	0	5	psig
28	.05	.15	LINR	0	10mV	Purity
29	0	250	LINR	0	5	psig
30	0	800	LINR	1	5	mmhg

M10	10 millivolt input:	Gain = 1000
M50	50 millivolt input:	Gain = 200
V1	1 volt input:	Gain = 10
V5	5 volts input:	Gain = 2
V10	10 volts input:	Gain = 1

11. Standard deviation analysis is run if necessary.

12. Finally, any remaining data that is being collected is stored.

The computer is then quiescent until the end of the 10-second period, at which time the cycle is repeated.

The second machine, Tweedledee, is used primarily for the writing and testing of any new software for Tweedledum. As Tweedledum and Tweedledee are configured identically, it should be possible to download Tweedledum with code that is bug-free. As a secondary function, Tweedledee could be used to do extensive analysis of any data gathered by the process machine. However, the time and manpower available is a function of the software development program for Tweedledum.

With a system in place measuring, supervising, and controlling the plant (with the operators and foreman made familiar with its features and trained in its operation), the initial phase of the project is complete. The cost estimates detailed earlier in this chapter expected 23 control loops to take about six months to close and to tune. This estimate is not based on substituting a PID algorithm in digital form within the machine for every control loop, but it is based on developing for each loop a many-branched decision tree, which may or may not interlock with other decision trees so that the machine is making many of the routine decisions formerly made by the operator. It may well be possible to include sequencing algorithms within the trees that will allow the computer to ramp the plant up and down to different production rates. If this could be done, then some very sophisticated alternatives become possible, such as economic optimization. However, it may take 12 months before the plant responds in a sufficiently predictable manner. Only then could the machine begin to attempt to coordinate control according to some economic criteria. As so much is involved and as plants vary so widely, there is no way to estimate a meaningful time frame as to when any one plant could actually be run under automatic economic optimization.

Two important areas that definitely affect the viability and the life of the installed computer systems must be considered. The first is technical obsolescence. The day the machine is delivered it is technically obsolete. No manufacturer will guarantee how long that machine you bought will be in production and for how many years spare parts will continue to be manufactured after production is halted. But no matter how many years it is in production, spare parts will inevitably become a problem. Worse yet, there are not enough competent maintenance men to go around, even today. So in the future, even if you were to find the parts, you may not find anyone capable of troubleshooting or installing them. The explosive growth of microelectronics can only exacerbate an already severe problem. Therefore, when installing the systems, bear in mind that everything on the computer side of the field-wiring termination panel is vulnerable and will likely be replaced once, if not twice, before the chemical plant reaches the end of its economic life.

The other important area is software and documentation. If the software is written in machine-specific form, then the usability of the software is tied to the availability of that specific kind of machine. It is better to develop software that is compiled into machine-compatible form in two stages. The first stage is a machine-independent pseudomachine code that in turn is very rapidly transformed into machine-specific code. Then, should the processor be changed for whatever reason, only the machine-specific code has to be regenerated from the pseudomachine code. It should not be necessary to rework the original source code. This also gives transportability to the source code and pseudomachine code. Offsite software development becomes feasible.

But more important than the direct problems of the programming languages, the individual's style, and the operating system in use are the indirect problems of support documentation. Without good documentation, the total software effort is to no avail, since the programs cannot be passed on to others. New employees then either have to waste time puzzling out what the original programmer did or be non-productive while reinventing the wheel. Documentation is emphasized elsewhere in this book, but it does bear repeating as so few heed the message: software is good only if it can be passed on, and only that which is properly documented is passed on.

7

PROCESS MEASUREMENT, MATHEMATICS, DATA LOGGING

Where to Start

The installation of a programmable processor and its associated peripheral equipment is a very expensive proposition. For the installation to be worthwhile, it must save money, increase the capacity of the plant, or improve the quality of the product. Each of these is a measurable result of the installation. But the extent of the improvements can be assessed only by comparing them with results previous to the installation. The immediate task of the installation, then, is to provide a base survey of the current operation of the plant. This must be done before any control work is attempted.

It is important to stress at the beginning of this chapter that the process operation improvement does not come from the mathematical accuracy of the processor. Indeed, the processor can carry calculations to 7 or 13 significant figures, but that is not the source of the improvement. All calculations are limited by the accuracy of the inputs. In process measurements all inputs but the most precise will be significant only to three figures. As the derived data cannot be more accurate than the data entered, then the output data also will be significant only to three figures. It must be a frequent exercise in process computer work to keep the results in accord with the accuracy of the data entered. Results must never be left at the number of significant figures to which they have been carried in the caculations. It is misleading. The one advantage slide rules have over electronic calculators is that they do not imply non-existent accuracy.

How do you prove what benefits can accrue to the process by installing a processor? The proof is the difference between the original operation of the process and the operation of the process under processor supervision and processor control. The benefit is the dollar value of these improvements. But before the survey can be made, before the accuracy of the results is questioned, and before the dollar benefits are reckoned, a start must be made, and we must begin at the beginning. As with all tools and machines, *doing* is the best teacher and the best experience. The making of the survey and the questioning of the accuracy of the results give a very firm foundation to the later control work.

Where is the beginning? The beginning is any one of those technical problems that currently beset the process. In the Witches' Brew Process six problems need solutions:

1. The distillation tower will not run beyond 85% of its nameplate capacity
2. The flash drum vents too much material
3. On occasions, the aqueous solute from the extractors carries away unacceptable quantities of the product
4. The solvent-to-solute ratio is very high, much higher than expected
5. Difficulty is encountered keeping Toil at its stoichiometric ratio to Bubble
6. The distillation tower's bottom flow freezes up frequently

Each individual solution will generate either a tangible dollar-saving or an intangible improvement in the process. Each individual problem has its own complexity and technical difficulties. By comparison of the effort and the reward, the problems can be ordered into a simple hierarchy from "most likely to save dollars" to "least likely to save dollars." For the Witches' Brew Process, the most attractive problem is the reactor stoichiometry. The programmable processor will be directed toward solving this problem. The machine's first task will be to collect, massage, and display whatever data on the problem can be culled from the process. This is the beginning.

As an aside to those who have never had association with chemical processes and programmable processors, the way information is presented is truly limited only by human imagination. Any inputs or any calculated variables can be presented numerically, plotted as curves in graphs, or given representation in bar charts. The information can be hard copy (on paper) or soft copy (video): in color, black-and-white, or gray scale; instantaneous or delayed, second-by-second, or a year at a time. The chapter on hardware describes in greater detail the physical limitations of the processor's peripheral equipment for storage quantities and for display dimensions. For the hardware associated with the Witches' Brew Process, our physical limitations are:

Soft-copy text	24 rows × 80 columns
Soft-copy plotting	256 dots × 512 dots
Hard-copy text	66 rows × 132 columns 88 rows × 132 columns
Hard-copy plotting	792 dots × 792 dots
Removable mass storage	64,000 real values 256,000 integers (each integer less than 255)

Input and Input Accuracy

What is the input data? In process measurement and control several different physical properties are used. Pressure and voltaic potential are two properties commonly used. The measurement is an analog of the actual physical property of interest (flow, level, temperature) and is not always a perfect indice of the true condition of the physical property. The devices that are used to take the measurements limit the knowledge of the physical property in three ways: (1) the device must not be exposed to extreme conditions such that it will be destroyed or made inoperable, (2) for meaningful resolution, the actual measuring range will be limited to a particular domain of interest, (3) although the physical property may exhibit a smooth continual increase from the low value to the high value, the measurement may not be linear.

The analog measurement correspondence with the actual physical property only exists within the ranged limits of the instrument. If the actual physical property is below the low limit of the instrument range, the instrument still will indicate the low-limit measurement. If the actual variable state is above the high limit of the instrument range, the instrument will indicate only the high-limit description. However, if the actual variable state is between the limits, then there is a one-to-one correspondence between the actual value of the state and the analog measurement. For any one measurement, there is one and only one state possible and, for any one state, there is one and only one measurement possible. The finer the resolution of the measurement, the greater will be the accuracy to which the measurement may be stated.

For the computer to ingest a measurement of any actual physical property, that measurement must be presented to the computer as an electrical analog, either voltage or current. Therefore, all pneumatic signals must be transduced to electrical analogs. In the computer interface that analog signal is changed to a numerical binary equivalent. It is this binary equivalent that the computer stores and uses as a representation of the original physical property. The binary equivalent is on a discrete integer scale; the exact range of that scale is peculiar to the particular computer and to the particular computer analog input interface used.

On an 11-bit binary scale, the range is (in decimal integers) from 0 to 2047, which is to say from (0) to (2^{11} - 1), or from 11 bits off to 11 bits on. On a 13-bit scale, the range is from 0 to 8191, which is to say from (0) to (2^{13} - 1), or from 13 bits off to 13 bits on. On a seven-bit binary scale the range is from 0 to 127, which is to say from (0) to (2^7 - 1), or from seven bits off to seven bits on.

The integer scale is a stairstep function with a multitude of discrete states. A one-to-one correspondence no longer exists between the physical property and the integer measurement stored within the computer. As an example, consider the temperature measurement of carbon dioxide gas (on a 0-100 degC range) by a pneumatic transmitter that transmits a 3-15 psig pneumatic analog signal. This pneumatic signal is transduced by a perfectly linear pneumatic-to-voltage transducer onto a 0V-5V (dc) signal. This 0V-5V (dc) signal is read by the computer onto an 11-bit integer scale. The correspondence of the scales in the cases of five temperatures is shown in Table 7-1. Because of the discrete nature of the integer scale, any temperature between 24.976 degC and 25.024 degC will be stored as 512 inside the computer. Also, note that 24.975 is stored as 511. At this point two factors should be emphasized.

First, it is pointless to employ measuring devices with measuring resolution greater than the input resolution of the computer. As pneumatic transmitters are considered to have a resolution of $\pm \frac{1}{2}\%$ of full scale for any measurement, and as an 11-bit integer scale has a resolution of $\pm 0.05\%$ of full scale for any input, this ordinarily should be no problem. Even the newer solid-state electrical transmitters have resolutions of only $\pm \frac{1}{4}\%$ full scale and again present no wasted resolution. No matter what the computer system and how many significant figures the machine can store and manipulate, the analog measurements must be read through the interface and that will be the determining constriction in the accuracy of all subsequent calculations.

Second, it is as well to note that for the finest input resolution, the analog input should be close to full scale. Table 7-2 displays a square root 0-10 division flow scale at the limits. The need for accurate ranging and zeroing of transmitters, transducers, and the interface—all elements in the chain—should be apparent. If the computer reads one instead of zero for "no flow," then it appears that 0.22 (almost a quarter) divisions of flow exist. If seven divisions is the normal rate of flow, then at no flow almost 3% of the normal flow will be registered. Mass balances calculated by the computer from small flows will hardly, if ever, balance.

Table 7-1
A Measured Variable, Two Corresponding
Analog Scales, and One Corresponding Integer Scale

Temperature (°C)	Pneumatic Signal (psig)	Electric Signal VDC	Integer Scale
0.0	3.0	0.00	0
25.0	6.0	1.25	512
50.0	9.0	2.50	1024
75.0	12.0	3.75	1536
100.0	15.0	5.00	2047

The conversion process from the continuous analog input scale to the stair-step integer scale can have a cumulative effect. Consider the mass flow calculation of carbon dioxide gas:

Mass flow = CF * DIV * OPF * CFC

where:

CF = actual flowing purity (analog scale linear 80-100%)

DIV = divisions flow (analog scale square root 0-10 flow divisions)

OPF = orifice plate factor (123.4567 lb/day division @90%, 50degC, 25psig)

CFC = correction for flowing conditions

$$= \frac{1}{CS} \sqrt{\frac{APF}{APS} * \frac{ATS}{ATF}}$$

where:

CS = purity at standard conditions, i.e., 90%

APF = absolute pressure flowing, i.e., (gauge pressure flowing + 14.696) psia. Gauge pressure flowing is analog scale 0-50 psig

APS = absolute pressure standard, i.e., 39.696 psia, i.e., 25 psig

ATF = absolute temperature flowing, i.e. (Centigrade temperature flowing + 273.2) degK. Centigrade temperature flowing is analog scale 0-100 degC.

ATS = absolute temperature standard, i.e., 323.2 degK, i.e., 50 degC

Table 7-2
The Problems of Digital Resolution

Integer Scale	Analog Input VDC	Actual Measurements Divisions Flow	Resolution
0	0.00000	0.0000	
			0.2209
1	0.00244	0.2209	
			0.0918
2	0.00489	0.3127	
↓	↓		
2045	4.99511	9.9951	
			0.0025
2046	4.99756	9.9976	
			0.0024
2047	5.00000	10.000	

In the following example and Table 7-3, all the flowing conditions are assumed to be midscale. The lowest possible analog and the highest possible analog are the range of possible measurements represented by the one discrete point on the integer scale. All measurements falling between 1023.500 and 1024.499 are covered by 1024. For the one set of 11-bit integer inputs in the table, two mass flows can be calculated: the lowest possible mass flow and the highest possible mass flow. The lowest mass flow occurs at lowest purity, lowest flow, lowest pressure, and highest temperature. The highest mass flow occurs at highest purity, highest flow, highest pressure, and lowest temperature.

$FLOW_{lowest}$ at flowing conditions of 7.0711 div, 90.0000%,
   50.0489 degC, 25.0000 psig:

$$FLOW_{lowest} = 90.0000*7.0711*123.4567$$
$$* \frac{1}{90.0000} * \sqrt{\frac{39.6960}{39.6960}} * \frac{323.2000}{323.2489}$$
$$= 872.9086 \text{ lb/day}$$

$FLOW_{highest}$ at flowing conditions of 7.0745 div, 90.0098%,
   50.0000 degC, 25.0244 psig:

Table 7-3
Digitizing Errors

	11-Bit Integer Scale	Lowest Possible Analog	Highest Possible Analog	Error % Digitizing
Purity flowing	1024	90.0000	90.0098	.011
Divisions flow	1024	7.0711	7.0745	.048
Temperature flowing degC	1024	50.000	50.0489	.098
Temperature flowing degK		323.2000	323.2489	.015
Pressure flowing psig	1024	25.0000	25.0244	.098
Pressure flowing psia		39.6960	39.7204	.061

$$FLOW_{highest} = 90.0098*7.0745*123.4567$$

$$*\frac{1}{90.0000}*\sqrt{\frac{39.7204}{39.6960}*\frac{323.2000}{323.2000}}$$

$$= 873.7578 \text{ lb/day}$$

$$FLOW_{difference} = 0.097\% \approx 0.1\%$$

Notice that the round off in the integer scale can introduce an error; in this case 0.1%.

For any product, the sum of the errors of the individual multiplicands equals the error in the product. In this case error in product equals the sum of the errors of the individual multiplicands.

Mass flow = Purity + Divflow + ½ (Pres + Temp)

For the digitizing errors,

0.097 = 0.011 + .048 + ½ (.061 + .015)

Half of the overall digitizing error comes from the division-flow round off. The offsets added to the purity, pressure, and temperature analog measurements in their individual instances reduce their digitizing error. The square root of the pressure and temperature measurement reduces their error even further. But there are also transmitter calibration errors and actual measurement errors. Even transmitters that have been spanned correctly and

Table 7-4
Transmitter Errors

| | Transmitted Analog | | Multiplicand Engineering Value | |
	Error	Range	Range	Error
Purity	±1/2%	20%	80%-100%	±0.10%
Divisions				
flow	±1/2%	(10 Div)²	0-10 Div	±0.25%
Temperature	±1/2%	100 degC	273-373 degK	±0.13%
Pressure	±1/2%	50 psig	14.7-64.7 psia	±0.39%

$$\text{Mass flow error} = 0.10 + 0.25 + \tfrac{1}{2}\,(0.13 + 0.39)$$
$$= 0.10 + 0.25 + 0.065 + 0.195$$
$$= \pm 0.61\%$$

zeroed are still subject to ±½% calibration errors. Table 7-4 shows that this can lead to a ±0.061% calibration error in the calculated mass flow. Measurement errors are even worse. Table 7-5 assumes various percentage errors in the repeatable resolution of basic measurements. Overall they lead to ±1.19% error in the mass flow. These errors are inherent in the basic nature of the various measurements and are not to be confused with transmitter repeatability calibration errors.

In any flow calculation with a computer the accuracy of the measurement of the flow has the greatest impact on the accuracy of the resulting calculated mass flow. Therefore, any effort spent to improve the overall accuracy must be spent on the orifice plate differential pressure measurement before any efforts are expended on temperature, pressure, or purity. In our example, even when all the analog inputs were assumed to be at midscale, there was a digitizing error of 0.1%. This means that solely on the basis of this error, the calculated answer can be given to only three significant figures; lowest mass flow 873 lb/day, highest mass flow 874 lb/day. But if we include the possible effects of transmitter calibration errors, the flow is subject to an error of ±6 lb/day. In addition, if we include the measurement errors, the discrepancy becomes ±17 lb/day. The true real-world flow lies somewhere between 856 lb/day and 890 lb/day. Almost three-quarters of that range are caused by errors associated with quantifying the orifice plate differential pressure measurements. The mass flow calculations can never be better than the measurement of the associated differential pressure.

We can now discuss the disease, "delusions of accuracy." It is the responsibility of the programmer, chemical engineer, and plant manager to ensure that no numbers calculated by the computer are printed or displayed with an implied level of accuracy beyond that which is possible. As an example, temperatures on a 0-100 degC scale should be displayed only to the nearest half-degree, i.e., 58.0 or 58.5, and should not be displayed as 58.378. Fur-

Table 7-5
Measurement Errors

	Repeatable Resolution	Percentage Error in Max Engineering Value
Purity	0.1%	± 0.1%
Divisions		
flow	0.1 Division	± 1.0% (Div)
Temperature	0.1 degC	± 0.035 degK
Pressure	0.1 psig	± 0.15% (psia)

Mass flow error = 0.1 + 1.0 + ½ (0.03 + 0.15)
= 0.1 + 1.0 + 0.015 + 0.075
= ± 1.19%

thermore, scientific notation has no place in the computer operation of a plant. 0.987654 E02 degC is a confusing way to write 98.7654 degC, which, as already mentioned, should be displayed as 99.0 degC. The misleadingly implied accuracy apart, each person who reads the number in scientific notation mentally converts it to 98.7654 anyway. So why not let the computer perform the mental drudgery? If the person formatting the display does not know the magnitude of the quantity involved, it is his job to calculate the range of possible values and to format the field accordingly in the familiar decimal number way—without an exponent.

Always keep the computer-displayed or computer-printed quantities in the same familiar units that are used in the everyday operation of the plant. Particularly avoid displaying absolute temperatures and absolute pressures. The project is difficult enough without antagonizing people with new systems of units.

Floating-Point Math Calculations

Floating-point math was described in Chapter 5, but the implications of making calculations in floating-point math were not explored. Computers are not the unerringly superaccurate machines many people assume them to be. There are two constraints within them that affect the level of accuracy desired in the computer calculations: time and space. Many machines offer two levels of floating math: single precision and double precision. The latter uses twice as many bits to define a number as does the former. Therefore, in any given storage space only half as many numbers can be handled, and the speed of the calculations is more than halved; but the level of significance is more than doubled. Computer control of process plants does not require the accuracy of double precision.

The precision of a number in floating-math notation is inherent in the number of bits used in the binary mantissa. The magnitude of a number in floating-math notation is prescribed in the number of bits used in the binary exponent (see Table 7-6). Within the machine itself (whether the calculating is done with a software package or with a floating-point processor), any mathematical operation is carried to extra levels of significance by using a mantissa that is at least four bits longer than the mantissa used to store actual values. The longer mantissa provides for proper round-off and normalizing such that the decimal level of significance is unimpaired.

What effect does a 23-bit mantissa have on computations? It can be in some cases that the use of sophisticated algorithms is pointless, or in other cases timeliness has a greater impact on the accuracy of the answer than the restrictive nature of the mantissa. At the end of this chapter, two examples are worked at length as illustrations.

The Significance of Output

To print or display any of the stored numbers at the correct significance, two options are open. Either the displayed number can be a rounded-off value or it can be a chopped-off value. In the former case 0.875 will be 0.9, rounded to one decimal place. In the latter case 0.875 will be 0.8, chopped to one decimal place. It is important that the people who view the output displays know whether the data is chopped or rounded. It is even more important that the input/output format routines give the correct answer whether it is a positive number or a negative number, chopped or rounded. The software conversion of integers to floating numbers or floating numbers to integers (positive or negative) must be by rounding and not chopping. All users of canned floating-point software are cautioned to find out whether such software rounds or chops negative-floating numbers in formatting and to test the integer/floating conversion routine and the floating/integer conversion routine for positive and, more particularly, negative numbers. A simple test of -1.51, -1.49, -0.51, -0.49, $+0.49$, $+0.51$, $+1.49$, $+1.51$ should prove the software's correctness.

The complementary problem of integerizing analog inputs is the generation of analog outputs from integer scales. Where programmable processors are used in chemical processes, the majority of the analog and digital outputs from the computer will be used to adjust either the position of the set point of controllers or the position of control valves. The usual scale for set points of analog controllers is linear from 0 to 1000. Any one of a number of voltage or current analog signal scales can be used between the analog controller and the computer. But whatever analog range is used, it will be stored in the computer as an integer scale. Quite possibly it will be a 10-bit integer scale

Table 7-6
Range and Accuracy of Numbers for Various Floating-Point Notations

Exponent # Bits	Exponent # Bits + Sign Bit	Decimal Number Range of Magnitude	
7	8	2^{-128}	2^{127}
11	12	2^{-2048}	2^{2047}

Mantissa # Bits	Mantissa # Bits + Sign Bit	Accuracy Decimal Digits
23	24	7
27	28	8
55	56	17

of 0-1023. Obviously, such an output is not infinitely variable. It has discrete steps. In control work this leads to two considerations.

First, what is the effect of the discrete stairstep nature of the output? Consider a flow controlled at 150,460 lb/day at 765 on the linear-output scale. The flow is proportional to the square root of the linear scale. Table 7-7 gives the values of the flows for the positions adjacent to 765. There is a difference of 98 lb/day between the output positions. Therefore, input measurement of precise resolution and flow-calculation algorithms of great precision have no impact beyond the fourth significant figure. If it is calculated that the flow is 150,443 lb/day and the flow required is 150,503 lb/day, the computer is unable to mandate a change. In this example control is possible only to the nearest 100 lb/day. The stairstep nature of the output forces the pattern of control into a series of discrete incremental rates.

Second, in pulse-type outputs fractions of a pulse cannot be sent. Flow-control algorithms have been written that contain alterable proportional and reset factors. But the capability to tune the calculation such that the pulse requirement calculated is changed from 0.170 to 0.193 pulses or changes from 2.78 to 3.01 pulses does not change the result. For the former in either case, no pulse is sent; for the latter in either case, the result is three pulses. The refinement is without practical value.

Judging Improvement Statistically

The problems discussed notwithstanding, how can improvement in control be judged? There are two ways: either an objective statistical analysis concentrating on some measure of dispersion, or a subjective inspection of computer-collected, computer-printed data. The writing and implementation of a statistical analysis program is straightforward. The measure of dispersion to use is the standard deviation. Inspection is a much more complex task and can be approached in almost any way.

The sample standard deviation of a group of sample values is the square root of the sample variance of that group of sample values. The sample variance is defined as the average of the squared deviations of the sample values from the sample mean:

$$M = \frac{\sum_{i=1}^{N} (X_i)}{N} = \frac{X_1 + X_2 \ldots X_i \ldots + N_N}{N}$$

$$(SD)^2 = \frac{\sum_{i=1}^{N} (X_i - M)^2}{N}$$

$$= \frac{(X_1 - M)^2 + (X_2 - M)^2 \ldots (X_i - M)^2 \ldots + (X_N - M)^2}{N}$$

where:

X_i = the ith sample value

N = number of samples

SD = standard deviation

M = sample mean

The sample mean is found by adding all the sample values together and dividing by the number of samples. The sample deviation is found by dividing the sum of all the squared deviations by the number of samples and taking the square root of the result. Each squared deviation is found by subtracting the sample mean from each individual sample value and squaring the result. By algebraic manipulation, it can be shown that

$$(SD)^2 = \frac{N \sum_{i=1}^{N} (X_i)^2 - \left(\sum_{i=1}^{N} X_i \right)^2}{N^2}$$

Table 7-7
Effect of Adjacent Control Valve Positions on the Controlled Mass Flow

Scale Position	Mass Flow lb/Day	Difference lb/Day
764	150362	
		98
765	150460	
		98
766	150558	

This is a much easier method of calculating the sample standard deviation. Only two running sums are accumulated: the running sum of the sample values and the running sum of the squares of the sample values. There is no need to store the individual sample values. An outline of a simple program follows that for a given sample size calculates the mean, the standard deviation, the largest individual sample value, the smallest individual sample value, and the range between the greatest and smallest individual sample values.

Two or more parallel process trains allow two or more control schemes to be tried simultaneously. Obviously, under such conditions, all uncontrolled variables that impact the process can be considered to affect all trains equally. If the statistical analysis sampling program is expanded to handle several groups of samples simultaneously and the control schemes under trial are switched from one train to another, then the best all-round control scheme will be found very rapidly without the need to monitor uncontrolled variables.

If the appropriate programs are written for the switching of control schemes between trains, for the sampling of values, for the printing of results, and for the tabulating of the control scheme rankings; then the whole analysis is automatic and can be run around the clock. The scheme that consistently has the smallest standard deviation, no matter on which train it is used, has the least dispersion. From that, the conclusion can be drawn that that scheme effects the tightest control.

Sometimes it may be found that two schemes have identical standard deviations and that there is little difference between the means. But sample value for sample value (and this is particularly important in the analysis of process streams), the sample value of one scheme is consistently above or below the corresponding sample value of the other scheme. It still can be inferred that the one scheme is superior or inferior. It is a legitimate statistical inference, given the consistent hierarchy. Such a test is properly called a *non-*

parametric rank order statistical test. Readers are referred to any elementary college textbook of statistics for further information.

SD Algorithm

Decision #1:   sample size (20, 64, 100?)
Decision #2:   sampling rate (every 10 seconds, once an hour?)

START,   (The following initializes the program):
HIVAL   = 0
LOVAL   = very large number
SIMSUM   = 0
SQRSUM   = 0
TOTNUM = 100 (Sample size of 100 chosen)
         N = 0

LOOP,   (The following is performed at the sampling rate):
       X = current instantaneous sample value
SIMSUM   = SIMSUM + X
SQRSUM   = SQRSUM + X*X

   IF X.LT.LOVAL   LOVAL = X
   IF X.GT.HIVAL   HIVAL = X

       N = N + 1

   If N.EQ. TOTNUM   GOTO   ANSWERS

END   (Temporary end; reentry will be at LOOP at next sample time.)

ANSWERS,   (Finish calculations, output results):

   MEAN = SIMSUM/TOTNUM
     SD = (SQRT(TOTNUM*SQRSUM − SIMSUM*
           SIMSUM))/TOTNUM
   RANGE = HIVAL − LOVAL

OUTPUT   SD, MEAN, HIVAL, LOVAL, RANGE,
           TOTNUM
FINISH   (algorithm is complete)

Judging Improvement by Inspection

To decide between control schemes by inspection requires that the computer store masses of data and print it out on demand, presenting the salient features with great clarity. Such an approach consumes vast amounts of mass storage and does require imagination and expertise in data presentation. As an example, consider collecting data on nine variables in 32-bit floating-point format every 10 seconds. In 19 hours, 45 minutes 255,960 bytes of information will have been collected. This is just short of the full capacity of one side of a single-density floppy disk. Sampling 158 variables once a minute will fill the floppy in six hours, 45 minutes. Obviously, some form of data condensation will stretch whatever storage is available. Two techniques are feasible.

First, the values can be mapped onto reduced scales. By balancing the possible range of a variable and the resolution required within that range, any variable can be mapped onto an eight-bit scale. The eight-bit scale has a maximum range of either 0-255 or $-128-+127$. The mapped-integer value is generated from the floating-point value of the variable by the suitable choices of a constant and a factor for *each* variable such that

$$INTEGER = (VARIABLE - CONSTANT) / FACTOR$$

The floating-point value is regenerated

$$VARIABLE = INTEGER*FACTOR + CONSTANT$$

With this technique, a floppy disk will hold a whole day's once-a-minute sampling of 158 variables.

A second condensation technique can be used when a variable has values that are constant over extended periods of time. Every variable value stored has associated with it a weight. That weight is the number of successive samplings for which the value did not change. When the data is exploded during retrieval, the variable value stored is generated in a repeated fashion equal to the weight.

The data for inspection can be presented either in tabular numerical form or graphed in sequential form, or even in combined graphical and tabular form. Whatever method is used, each individual sheet of output must be headed, timed, and dated. Failure to do so makes the data meaningless. Tabular or columnar presentation of data requires no detailed explanation. Blocking is encouraged. Do use occasional blank rows and multiple columns of blanks so that the material can be scanned better by the eye. A page solid with numbers is undecipherable.

REACTOR #1	03/11/80	12:00	REACTANT "A"
			COOL WATER "B"

lb/DAY				gpm		
130			A	40		B
130			A	39	B	
130			A	40		B
125		A		39	B	
120	A			38	B	
98	A			40		B
99	A			38	B	
140			A	39	B	
140			A	39	B	

Figure 7-1. Simple Teletypewriter Plotting.

Material can be presented in graphical form even on the simplest of teletypewriters. Individual alphabetical characters are used to represent the plotted values of the variable. Figure 7-1 is an example of a mass flow whose value fluctuates between 98 and 140 lb/day and a volume flow hovering around 39 gpm. "A" represents the mass flow and "B" represents the volume flow. The values of the mass flow between 100 and 150 lb/day are plotted between columns 5 and 29 inclusive. For a flow of 130 lb/day, the column in which "A" is to be printed is calculated:

$$\text{Column number} = \frac{(130 - 100)}{(150 - 100)} * (29 - 5 + 1) + 4.5$$
$$= 19$$

It is easy to arrange that "A" be printed in column 19. Tidy programming will ensure that for values below 100 the "A" still appears in column 5 and that for values over 150 the "A" appears in column 29.

Several manufacturers of matrix printers offer true plotting capability. The computer can address any one dot position in a horizontal row, and the vertical resolution is to the individual dot line and not to the usual character row. The inclusion of one certain control character in any string of characters causes the printer to enter the plot mode. That string of characters will appear as the corresponding sequence of dots in but one dot line, and not as the usual alphanumeric characters in a character row. Resolution over standard 11 $\times$ 14½ inch computer form is 792 $\times$ 792 dots. As the switch between plot and print is software-controlled, one page can be a mix of plotting and the usual printing.

Such matrix printers can be used not only to present data in graphical form for inspection, but also can be used as multipoint recorder charts in real time. Plotting values every 10 seconds over a two-hour period requires 720 dot lines. Twelve dot lines per character row, a blank-character row at the top of the page, a blank-character row at the bottom of the page, and four character rows for title, date, time, and headings requires a total of 72 dot lines. The complete two-hour plot could be margined, titled, and headed on one 11-inch form. Across the width of the page, many combinations are possible: two plots each 360 dots wide, or even six simultaneous plots each 128 dots wide. In the latter case if for each plot one variable is represented by a single dot and another variable is represented by two adjacent dots, 12 different variables can be plotted in easily distinguished form.

There are four major advantages (and a host of minor advantages) in using a matrix printer in the plot mode as a multipoint recorder: (1) signals from dissimilar sources (millivolt thermocouples, pneumatic pressure transmitters, electronic flow transmitters) are all printed on the very same line simultaneously, (2) there is no offset between the individual records as there is on a multipen strip recorder, (3) the records do not run at different speeds. Any engineer who has tried to match two or more charts that run at different speeds in different parts of the panel board will appreciate this advantage, (4) which variables are plotted is easy to change and can be done by keyboard entry in a matter of seconds.

As there is only one paper source to check, only one ribbon to ink, and no pens to adjust, lost or spoiled records from torn chart papers and overinked/ underinked pens are negligible. Because simultaneous events are plotted in one line, residence times and delay times can be measured. Even correlation becomes precise. For example, should the raw material feed in a reactor fall off *before* the pressure increase by as little as 10 seconds, the hard copy will carry the incontrovertible proof of the relative timing of the two events. Each record carries the correct time and date. The plotting scales can be changed rapidly. At the expense of range, the resolution of the plot of any variable can be magnified greatly, revealing hitherto unsuspected variations. Finally, the printed dot is frictionless and weightless. Inking pens have three mechanical problems that prevent them from ever giving a precise indication: the friction of the pen against the paper, the hysteresis of the bellows or galvanometer, and the backlash and friction of the levers and couplings within the recorder.

The computer capability of real-time sampling, data storage, very fast arithmetic, and report generation make it an invaluable tool. The rest of this chapter describes a variety of ways of presenting process information.

Presenting Process Information

Unlike conventional instrumentation, a programmable processor can present its measurements in a variety of ways, each available at the touch of a key. Further, it can manipulate its measurements and present not plain input data but derived data, summaries, or enhancement. The limit is the limit of the imagination of the enquiring mind. The very concepts that will allow the programmable processor to be used as process controller or process supervisor also allow it to be used as a process microscope, process electrocardiogram, or process x-ray machine. The programmable processor will report and investigate—punctually, reliably, and continually—anything associated with the process if the relevant data is presented to the front-end. Imaginative collection and presentation of process information lead to process insights, improved process knowledge, and ultimately to highly efficacious process control.

Eight formats presenting process information are introduced in this chapter. (The specific use of each format is given in Chapter 8):

Format 1

Single-input real-time checking (soft or hard copy) [CRT; line printer]. Presents numerical information about any one individual analog input.

Format 2

Triple-variable real-time tracking (soft or hard copy) [CRT; line printer]. Presents information on any three variables as numbers and plots.

Format 3

Multiple-variable real-time plotting (hard copy) [matrix printer]. Presents plots on up to 20 variables simultaneously.

Format 4

Multiple-variable real-time tracking (hard copy) [line printer]. Presents information on up to 20 variables as numbers, plots, or numbers and plots.

Format 5

Multiple-variable batch plotting (hard copies) [matrix printer; floppy]. Presents plots every shift on 120 variables in six sets of 20 plots.

Format 6

Multiple-variable batch tracking (hard copies) [line printer; floppy]. Presents information on up to 32 variables in two exclusive sets of 16 as numbers, plots, or numbers and plots (tracked at operator-chosen intervals).

Format 7

Individual variable investigation (soft copy, hard copy) [line printer; CRT]. Calculates mean, standard deviation, and analyses changes between successive values of any chosen variable.

Format 8

Multiple-variable investigation (soft or hard copy) [CRT; line printer]. Calculates mean, standard deviation, and analyses changes between successive values of 16 operator-chosen combinations of variable, interval, and time of day.

Format 1

Title.   Single-Input Real-Time Checking (Soft Copy) (Numbers Only)

Objective.   To present numerical information about any one individual analog input

Output example.   Figure 7-2

Description.   Every cycle fresh information is entered on the bottom line of the display, causing the information already on the screen to scroll upward one line. The top line (containing the oldest information) is lost. Each line contains the following fields, from left to right:

Field 1.   Index number of the analog input being tracked.

Field 2.   Analog measurement input voltage as applied to the screw terminals.

Field 3.   mV for millivolts; V for volts.

Field 4.   Processor's internal integer value of the input voltage.

Field 5.   Percent of scale of analog measurement input voltage signal.

Field 6.   Actual instantaneous engineering value of input.

Field 7.   Units, name, and brief description of analog input.

Use.   Request the display, and if a different input is to be tracked, the index number of the input desired is entered by request through the screen and keyboard, bottom line, left-handmost field. Fresh information enters the complete bottom line every cycle, scrolling previous information upward one line.

```
 75     1.040 V   0652   20.81%    1.018
 75     1.040 V   0652   20.81%    1.018
 75     1.040 V   0652   20.81%    1.018
 75     1.040 V   0652   20.81%    1.018
 75     1.042 V   0653   20.85%    1.048
 75     1.824 V   1353   36.49%    4.542
 75     2.090 V   1530   41.81%    5.223    DIV MARES SWEAT FEED
 75     2.188 V   1600   43.77%    5.452
 75     2.154 V   1562   43.08%    5.373
 75     2.191 V   1601   43.82%    5.458
 75     2.193 V   1602   43.86%    5.463
226     1.746 V   1313   34.92%  174.787    PSIG SOAKER
226     1.744 V   1312   34.88%  174.641
226     1.741 V   1311   34.83%  174.494
226     1.739 V   1310   34.78%  174.347
 40    10.000 mv  3777  100.00%  213.930    DEGC BUBBLE STORAGE TANK
 39     8.343 mv  3254   83.43%  182.385    DEGC DC REBOILER OUT
 39     8.319 mv  3247   83.19%  181.947
 39     8.309 mv  3245   83.09%  181.761
 39     8.324 mv  3250   83.24%  182.040
234     2.181 V   1575   43.62%   43.624    LEVEL EXTRACTOR DRUM #2
234     2.178 V   1574   43.57%   43.575
234     2.203 V   1606   44.06%   44.064
234     2.188 V   1600   43.77%   43.771
234     2.188 V   1600   43.77%   43.771
```

Figure 7-2. Single-variable real-time checking of analog inputs (soft copy).

Coding.   The entered index number is stored as the key variable for this display. At display generation, it is used as the offset into the various tables to find the AI gain, the digital value, the engineering value, and the descriptive label of the input. Subsequently, the input voltage and the percent of scale are calculated. The various values then are formatted into a display string and written to the device presenting the display. Care is taken to scroll the screen first by writing either a blank in the bottom right-hand corner or a line feed in the bottom left-hand corner. The descriptive label is written the first cycle after a new index number is entered, and every 12 cycles thereafter.

Comments.   Hard-copy tracking is an easy modification. Send the display string to a printer, and follow it with a carriage return and line feed. To stop the hard-copy tracking, return the printer to its inactive state.

Format 2

Title.   Triple-Variable Real-Time Tracking (Soft Copy) (Numbers and Plot)

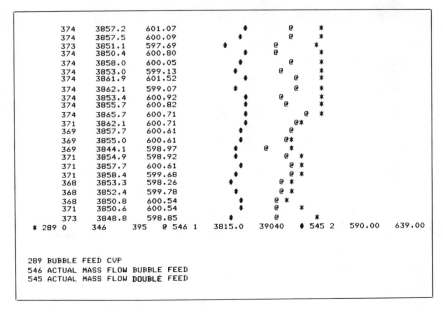

Figure 7-3. TRACKING EXAMPLE—triple input; real-time tracking; sort copy.

Objective. To present information about the engineering values of any three variables both as numbers and as mapped plots.

Output example.   Figure 7-3 (soft copy)
                     Figure 7-4 (hard copy)

Description. The display is 23 lines of a 23-cycle history (aged upward a new line every refresh cycle) and one line of 12 operator-entered values. For each line of history, the fields are left to right:

Field 1. The engineering value of the first variable.

Field 2. The engineering value of the second variable.

Field 3. The engineering value of the third variable.

Field 4. A 50-column wide plot into which each of three variables has been mapped between individual pairs of limits chosen by the operator.

The bottom line consists of 15 fields left to right:

Field 1. An asterisk (*) mapping symbol of variable one.

Field 2. The index number of variable one.

Field 3.   Number of decimal places.

Field 4.   The lower limit for the mapping of variable one.

Field 5.   The upper limit for the mapping of variable one.

Field 6.   A commercial "at" sign (@) mapping symbol variable two.

Field 7.   The index number of variable two.

Field 8.   Number of decimal places.

Field 9.   The lower limit for the mapping of variable two.

Field 10.   The upper limit for the mapping of variable two.

Field 11.   A "pound" sign (#) mapping symbol variable three.

Field 12.   The index number of variable three.

Field 13.   Number of decimal places.

Field 14.   The lower limit for the mapping of variable three.

Field 15.   The upper limit for the mapping of variable three.

Use.   Request the display and change any of the operator-entered values as required by positioning the cursor and then entering the new value with the screen and keyboard routine. Each of the three index numbers, each of the three decimal-place options, and each of the six limits are operator entries. At each refresh, the new data is written to line 24, the complete display scrolled up one line, and the line of operator-entered values restored to line 24.

Coding.   The entered index numbers are used as offsets into the engineering values table to find the required values. For each of the three variables in turn, the values are mapped onto a 50-count integer scale using the appropriate limits. Any value less than or equal to the lower limit is accorded the value one, and any value equal to or greater than the upper limit is accorded the value 50. The integer value of the mapping in turn becomes the column position offset for the associated plot character between columns 31 and 80. Writing the display string to the bottom line of the device presenting this display will overlay the operator-entered values and scroll the screen upward, leaving a blank bottom line ready to receive a rewrite of the line of operator-entered values.

Comments.   Hard-copy tracking is an easy modification. The greater columnar width available allows a greater-count integer scale for greater mapping resolution.

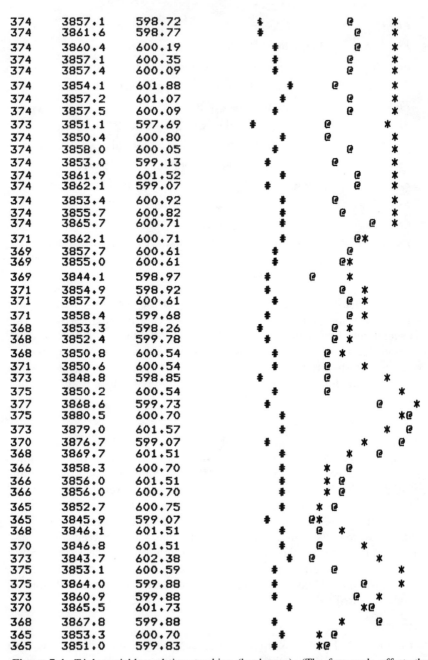

374	3857.1	598.72
374	3861.6	598.77
374	3860.4	600.19
374	3857.1	600.35
374	3857.4	600.09
374	3854.1	601.88
374	3857.2	601.07
374	3857.5	600.09
373	3851.1	597.69
374	3850.4	600.80
374	3858.0	600.05
374	3853.0	599.13
374	3861.9	601.52
374	3862.1	599.07
374	3853.4	600.92
374	3855.7	600.82
374	3865.7	600.71
371	3862.1	600.71
369	3857.7	600.61
369	3855.0	600.61
369	3844.1	598.97
371	3854.9	598.92
371	3857.7	600.61
371	3858.4	599.68
368	3853.3	598.26
368	3852.4	599.78
368	3850.8	600.54
371	3850.6	600.54
373	3848.8	598.85
375	3850.2	600.54
377	3868.6	599.73
375	3880.5	600.70
373	3879.0	601.57
370	3876.7	599.07
368	3869.7	601.51
366	3858.3	600.70
366	3856.0	601.51
366	3856.0	600.70
365	3852.7	600.75
365	3845.9	599.07
368	3846.1	601.51
370	3846.8	601.51
373	3843.7	602.38
375	3853.1	600.59
375	3864.0	599.88
373	3860.9	599.88
370	3865.5	601.73
368	3867.8	599.88
365	3853.3	600.70
365	3851.0	599.83

Figure 7-4. Triple-variable real-time tracking (hard copy). (The four-cycle offset, the sensitivity of the system to valve movement, and the overall close correlation between the (*) valve position and the (@) corresponding mass flow are particularly noticeable.)

Figure 7-4 continued

368	3848.8	600.58	‡			@	✳			
370	3845.7	601.46	‡		@			✳		
373	3849.4	600.70	‡			@			✳	
375	3850.2	599.17	‡			@				✳
377	3858.1	603.19		‡			@			@ ✳
377	3872.6	603.19		‡					@	✳
375	3871.5	600.80	‡						@ ✳	
373	3877.4	600.70	‡					✳	@	
370	3860.5	600.70	‡				@✳			
368	3855.2	599.78	‡			@				
368	3852.0	601.36	‡		@✳					
370	3860.7	602.17		‡		@✳				
370	3852.4	601.25	‡		@	✳				
373	3857.7	600.38	‡			@	✳			
373	3852.9	599.67	‡		@		✳			
373	3852.5	599.56	‡		@		✳			

Format 3

Title.   Multiple-Variable Real-Time Plotting (Soft-Copy Set-Up) (Hard-Copy Plots)

Objective.   To present continual real-time hard-copy plots of up to 20 variables with easy changing of any of the variables plotted or the individual plot limits; each page of plots to be titled correctly.

Output example.   Figure 7-5 (soft-copy setup)
                             Figure 7-6 (hard copy)

Description   (soft-copy setup).   This is 20 lines of operator-entered input indicating which variables are to be mapped and the limits to the mapping. Each line is left to right:

Field 1.   Index number of required variable.

Field 2.   Number of decimal places for the limits and actual value in hard-copy heading.

Field 3.   Lower limit for the mapping of the variable.

Field 4.   Current actual value of variable.

545	1	200. 0	625. 4	1000. 0	DOUBLE MASS FLOW #/H
546	1	20000. 0	31256. 7	50000. 0	BUBBLE MASS FLOW #/H
5	1	100. 0	121. 1	190. 0	DEGC SOAKER IN
8	1	100. 0	190. 9	190. 0	DEGC SOAKER OUT
553		0	123	800	GPM COOLER WATER
297		750	876	1000	CVP COOLER WATER
11	1	50. 0	49. 8	100. 0	DEGC FLASH DRUM EXIT
555	1	250. 0	375. 6	500. 0	MARES SWEAT MASSFLOW #/H
556	1	0. 0	125. 3	500. 0	GAZ MASSFLOW #/H
558	1	17000. 0	17017. 1	20000. 0	FROG SPIT MASSFLOW #/H
559	1	17. 0	17. 1	20. 0	RAFFINATE VOL FLOW GPM
100	1	0. 0	70. 0	100. 0	SOLVENT TANK LEVEL %
18	1	20. 0	29. 5	40. 0	EVAPORATOR TOP DEGC
565	1	0. 0	5130. 6	30000. 0	EVAPORATOR STM FLOW #/H
20	1	75. 0	90. 5	100. 0	EVAPORATOR BTMS DEGC
569		0	1250	1500	DISCOL FORWARD FLOW #/H
570		0	39923	90000	DISCOL REFLUX FLOW #/H
571		0	351	2000	DISCOL CONDENSER WTR GPM
					Y

Figure 7-5. Multiple-variable real-time plotting (CRT setup, soft copy).

Field 5.   Upper limit for the mapping of the variable.

Field 6.   Variable description for hard-copy heading.

The display is in five groups of four, in the same order as the hard copy, four plots per group, five groups on a page. The final field in the bottom right-hand corner starts/stops the plotting.

Description (hard-copy output). There are eight lines of alphanumeric heading and 720 dot rows of plotting per page. Above each group of plots are up to four titles and 12 numbers. There is a title for each plot in the group. For each plot beneath the title there are three numbers in a line: the mapping low limit, the actual value of the variable at the beginning of the plot, and the mapping high limit. One dot row is printed per cycle. The groups are separated by solid bands regularly broken to mark the passage of time. The date and time are superimposed on the eighth line of the heading at the far right-hand side.

Use.   Request the soft-copy setup. Change any index numbers, decimal places, limits, and titles as required with cursor positioning and entry by request through the screen and keyboard. When all the changes are complete,

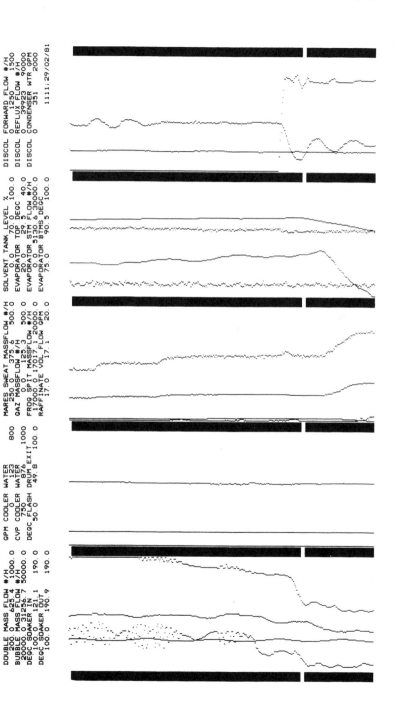

Figure 7-6. Multiple-variable real-time plotting (hard copy).

move the cursor to the final field, and enter a "Y" to begin plotting. To stop the plotting, enter an "N."

Coding.   For the plotting, the entered index numbers are used as offsets into the engineering values table to find the required values. Each value is mapped into a 142-count integer scale between 2 and 143 using the appropriate limits. Excessive values are mapped to 2 or 143. Finally, using the group number as a further offset, the mapped integer value is used to position the plot dot. To prevent excessive values mapped at the limit from being confused with the solid demarcation bands, the 1st and 144th dots of the 24-column plot are left intentionally blank. Six bands two columns wide (12 dots) and five plots 24 columns wide (144 dots) total 132 columns (792 dots). The heading occurs when the plot is initiated, and for every new page thereafter. The information in the heading is taken from the soft-copy setup. Some extra flexibility and a faster setup can be obtained by using the following protocol: If INDEX NUMBER=LOW LIMIT=0 (whatever the value of the high limit), there is no variable or constant to be plotted. If INDEX NUMBER>0, but LOW LIMIT=HIGH LIMIT=O, then map the variable using the limits assigned the previous variable. If the INDEX NUMBER and the HIGH LIMIT = 0 (but the LOW LIMIT is not 0), then map the low-limit number as a constant, using the limits assigned the previous variable. In all other cases where the index number is greater than zero, and the high limit is not zero, then map the variable according to the limits.

Comments.   The plotting can be done only on those matrix printers that permit individual dot rows to be printed under software control. If the printer/plotter is not used exclusively for plotting, then uninterrupted plots cannot be guaranteed. Five groups of four plots at 142-dot resolution is a personal choice by the author. It gives sufficient clarity, reasonable resolution, and sufficient variables to investigate any one problem.

Format 4

Title.   Multiple-Variable Real-Time Tracking (Soft-Copy Setup) (Hard-Copy Numbers and Plots)

Objective.   To present information about the engineering values of up to 20 variables simultaneously, both as numbers and as mapped plots, the formatting of the hard-copy output to be set as the variables are chosen.

Output example. Figure 7-7 (soft-copy setup)
          Figure 7-8 (hard copy)

```
0837 02/29/81     GROUPS 6    KIND OF COPY 5   EVERY 6
655 20   1 6.1          A                3684.0          BUBBLE TARGET FLOW #/H
288 22     4.0  23  10  B     270         275     279    BUBBLE CVP
 65 22     5.2  33  20  C     8.40        8.47    8.59   BUBBLE DIVISIONS FLOW
545 22     6.1  33  20  D   3655.0      3683.7  3702.5   BUBBLE ACTUAL FLOW #/H
326 44 115 6.1  65  50  E    650.0       658.9   699.0   TOIL ACTUAL FLOW #/H

290 44     4.0  65  50  F     610         636     659    TOIL CVP
 67 44     5.2  65  50  G     7.81        8.00    8.30   TOIL DIVISIONS FLOW
331 10   1 8.1          H                1249.9          EXTFEED EPR #/H
332 10  17 8.1          I                3030.1          EXTFEED WITCHES BREW #/H
340 10  51 8.1          J                7676.5          EXTFEED WATER #/H

333 10  67 7.1          K                875.3           EXTFEED TROUBLE #/H
334 10  81 7.1          L                369.8           EXTFEED HEAVIES #/H
558 10  33 9.1          M               17456.7          EXT FROG SPIT FEED #/H
343 10   9 8.1          N                1258.1          EXTRACT EPR #/H
344 10  25 8.1          O                3333.3          EXTRACT WITCHES BREW #/H

345 10  42 9.1          P               17654.3          EXTRACT FROG SPIT #/H
341 10  59 8.1          Q                7586.4          RAFFINATE WATER #/H
336 10  74 7.1          R                864.2           RAFFINATE TROUBLE #/H
337 10  88 7.1          S                381.6           RAFFINATE HEAVIES #/H
                        T
```

Figure 7-7. Multiple-variable real-time tracking (CRT setup, soft copy).

Description (*soft-copy setup*). This is 21 individual lines of operator-entered input that is used to collect and format the hard-copy output. The first line is (left to right):

Field 1. Military time.

Field 2. Month/day/year.

Field 3. Group or groups in output (groups are four, two and one. A seven will output all groups because $4+2+1=7$).

Field 4. 1 = *no copy*
2 = *hard copy*
3 = *hard copy plus time each line*
4 = *hard copy plus time and date each line*
5 = *paged hard copy.*

Field 5. Number of cycles between output of lines.

The remaining 20 lines are (left to right):

Field 1. Index number of required variable.

Field 2. *1st digit* = variable to be printed if this group is selected
2nd digit = variable to be plotted if this group is selected.

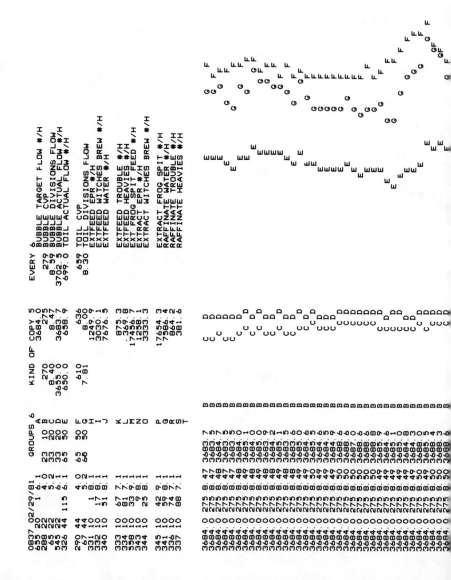

Figure 7-8. Multiple-variable real-time tracking (hard copy).

Field   3.   Initial column for numerical output of variable.

Field   4.   *1st digit* = overall columnar width for numerical
output of variable
2nd digit = number of decimal places.

Field   5.   Initial column for mapped output of variable.

Field   6.   Columnar width for mapped output of variable.

Field   7.   Letter used to represent mapped value of variable.

Field   8.   Lower limit for the mapping of the variable.

Field   9.   Current instantaneous actual value of variable.

Field 10.   Upper limit for the mapping of the variable.

Field 11.   Units, name, and brief description of variable.

(*Hard-copy output*). According to which of the three groups are required as the hard-copy output. It will be formatted according to the directions included in the soft-copy setup. Figure 7-8 is a "paged hard copy" of groups "2" and "4" every six cycles. A copy of the setup is included with every 60 lines of output for future reference purposes.

Use.   Request the display. If any of the operator-entered values require changing, position the cursor, and enter the new value with the usual routine. On any one line, any of the following can be changed: index number, print group, plot group, initial print column, number width, number of decimal places, initial mapping column, map width, lower map limit and upper map limit. Finally, where necessary, make the changes in the top line: group or groups to form the output, the frequency of the output, and whether or not the output is preceded by the time and by the date.

Coding.   The entered index number is used as the offset to obtain the current engineering value of the variable. If the individual print group is of the required output group, the number is formatted to the correct width and number of decimal places at the correct column. A zero value entered at initial column will cause the field to be formatted adjacent to the field formatted immediately previous. Also, a zero width and zero decimal places will cause the immediately previous format to be reused. If the individual plot group is of the required output group, then the engineering value is mapped into the defined columnar space only if it is between the upper and lower limits; otherwise, extreme values are plotted at the map limits. The given alphabetic character is used to mark the mapped position. When all 20 possible requests have been checked, and where necessary converted to numerical and mapped values, the display string is sent to the printing device.

As an element of sophistication, the entered number of decimal places can be used to reformat the lower limit, the instantaneous value, and the upper limit on the soft-copy setup. Also, the entered index number is used to obtain the description from the descriptions table for each of the variables in the soft-copy setup.

Comments.   The hard copy is of archival quality only if it carries the time, the date, and a description sufficient to explain the output. Without all three, the hard copy is of immediate interest only and has no long-term merit. The flexibility of choosing between all three groups, any two, just one or more, and having flexibility between number only, mapping only, and both number and mapping allows the one soft-copy setup to give very rapid access to different hard-copy printouts without excessive changes to the soft-copy setup. Unfortunately, real-time tracking at infrequent intervals may not give an uninterrupted printout on a printer that is not dedicated to the purpose.

Format 5

Title.   Multiple-Variable Batch Plotting (Soft-Copy Setup) (Floppy-Disk Mass Storage) (Hard-Copy Plots)
Note: a floppy disk is recorded a sector at a time. The length of a sector is usually 128 bytes (single density), and there are 2002 sectors to one side of a disk.

Objective.   Over a period of time to collect data that will subsequently be plotted in one burst. To maintain compatibility with the real-time plotting formats, and to use the storage capacity of a sector to its greatest advantage, 120 variables in six sections of 20 each will be collected simultaneously and stored as one floppy-disk sector. The data will be collected and stored every fourth cycle (every 40 seconds) and plotted every eight hours, when 720 collections will have been made. The fully constituted output will consist of six individually titled pages, one for each section, each page containing the plots of 20 variables in five groups of four.

Output.   Figure 7-9 (soft-copy setup)
Figure 7-10 (hard-copy output)

Description (soft-copy setup).   But for the addition of two extra fields in the top line, this setup is identical to the real-time plotting setup. One extra field is to request which of the six sections appears in the soft-copy setup. The other field is to request an up-to-the-minute plot of the currently displayed section.

92	1	0. 0	4. 3	10. 0	DISCOL STEAM DIV
111	1	100. 0	155. 5	200. 0	DISCOL STEAM PSIG
87	1	0. 0	7. 1	10. 0	DISCOL FEED DIV
238	1	90. 0	92. 6	100. 0	DISCOL FEED DEGC
108		700	770	900	DISCOL TOP PRES MMHG
109			850		DISCOL BTM PRES MMHG
88	1	0. 0	9. 8	10. 0	DISCOL BTMS DIV
241	1	0. 0	85. 0	100. 0	DISCOL BTMS % LEVEL
90	1	0. 0	3. 9	10. 0	DISCOL REFLUX DIV
89	1	0. 0	0. 0	10. 0	DISCOL FORWARD DIV
91	1	0. 0	1. 1	10. 0	DISCOL COND WTR DIV
242	1	0. 0	44. 4	100. 0	DISCOL REFLUX DRUM % LVL
21	1	90. 0	100. 1	150. 0	DISCOL OVERHEAD DEGC
26	1		107. 2		DISCOL TRAY #3 DEGC
29	1		114. 3		DISCOL TRAY #6 DEGC
32	1		120. 4		DISCOL TRAY #9 DEGC
34	1	90. 0	127. 9	150. 0	DISCOL TRAY #11 DEGC
36	1		129. 7		DISCOL TRAY #13 DEGC
38	1		130. 8		DISCOL TRAY #15 DEGC
244	1		131. 3		DISCOL BOTTOMS DEGC

Figure 7-9. Multiple-variable plotting (batch) (CRT setup, soft copy).

(Hard-copy output). This output is identical in format to the real-time plotting. Of course, one dot row now represents four cycles. The breaks in the solid dividing bands mark the hours.

Use. Six pages of output will occur at the end of each shift. If required, the last eight hours of any section will be plotted on request at any time.

Coding. Every fourth cycle the required variable values are mapped onto their individual 142-count integer scales. It is these 120 mapped values that are stored (one mapped value per byte for the first 120 bytes) in each sector. The absolute position of the sector used at any one time within the 720 sectors is governed by the absolute time of the collection cycle. At plotting, the sectors are read successively from oldest to most recent according to the absolute time of the initial plot. For each plotted dot row, the whole sector must be read, but only the particular 20 values of the section of interest are used to position dots. The format of the hard copy is identical to the real-time hard-copy plotting.

Comments. Obviously, this information generation required a printer plotter. But any bulk-storage medium can be used. The advantage of a re-

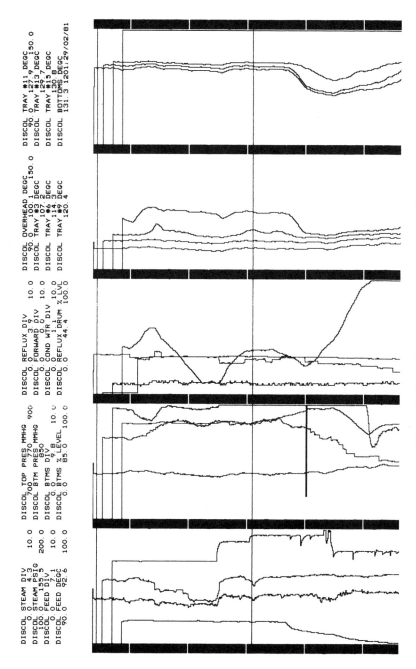

Figure 7-10. Multiple-variable plotting (batch) (hard copy).

movable bulk-storage medium is that it can be stored away from the processor and, if necessary, the data can be reconstituted by a second machine. At such a juncture, mix and match of plots from different shifts, or at different times, or from different sections could be made. With sufficient bulk storage, hourly, daily, monthly, and yearly data can be accumulated. Although the storage of the data as integer counts does give compaction, it has the disadvantages that the variable it represents and the limits that are used in the mapping are not stored. Consequently, the conformity of the integer scale over a period of time cannot be guaranteed. The longer the elapsed period of the data assembly, the greater the likelihood that the setup will have been altered. If the compressed data is to be stored for regeneration in the future, then the corresponding setup must be included with the complete block of compressed data.

The automatic generation of paper reports carries the inherent danger of producing information that is beyond comprehension, or more paper than can be read by any one person. Before data is gathered, it should be decided what is to be gathered, how often, its purpose, and how long the data and information should be kept before being erased or destroyed. The paper on which a report appears is inexpensive, but the binder, the folder, the file cabinet, and the office space which contain that paper are expensive. In fact, hard-copy attitudes decide whether the programmable processor is a tool or a burden.

Format 6

Title.   Multiple-Variable Batch Tracking (Soft-Copy Setup) (Floppy-Disk Mass Storage) (Hard-Copy Numbers and Plots)

Objective.   Over a period of time to collect data that subsequently will be output in one burst. A burst will be one page of output containing up to a maximum of 88 lines of information inclusive of the heading. Each sector of the floppy disk will store 32 numbers of 32-bit floating math. Each sector will be considered to be two sections of 16 numbers, gathered and stored simultaneously once a minute. The precise format and nature of the output will be governed by the soft-copy setup at the time of printing. The outputs are available as one burst on request, as continual bursts once requested, and at set hours of the day. The periodicity of the data in the burst is also alterable.

Output example.   Figure 7-11 (soft-copy setup)
Figure 7-12 (hard-copy output)

```
SECTION 1
#1 PH CHECKING                      0  0  0    0 N      0  0   0 N     1 10   1 60    6 Y
#2 DAILY DISCOL OPERATION 0  0  0   0 N      2 60 144  Y     0  0   0  0    0 N
#3                                  0  0  0    0 N      0  0   0 N     0  0   0  0    0 N

555 11   1 6. 1  60  20 A    300. 0       327. 9       360. 0 MARES SWEAT ACTUAL #/H
349 11     7. 1  60  20 B   3040. 0      3053. 3      3080. 0 WB TO NEUTRALISER #/H
298 11     4. 0  25  30 C     575          595          605 FLASH DRUM FWD FLOW CVP
231 11     6. 2  25  15 D    10. 55       10. 96       11. 14 NEUTRALIZER PH

567 02           5   20 E   3600. 0      4000. 8      4400. 0 DISCOL FEED #/H
311 20   1 4. 0                             617               DISCOL FEED CVP
570 02          29   20 G    80. 00       94. 29      120. 00 DISCOL REFLUX GPM
314 20  25 4. 0             H                362               DISCOL REFLUX CVP

569 02          53   20 I   1100. 00     1250. 73     1400. 00 DISCOL TOP PRODUCT #/H
313 20  49 4. 0             J                471               DISCOL TOP PRODUCT CVP
572 02          77   20 K   4500. 0      4964. 0      5500. 0 DISCOL STEAM #/H
316 20  73 4. 0             L                536               DISCOL STEAM CVP

244 20  97 6. 1            M                128. 4            DISCOL BTM DEGC
373 20 103 6. 2            N                0. 17             ANALYSIS EPR BTM PRODUCT
568 02         113   20 O   3000. 0      3210. 2      3400. 0 DISCOL BTM PRODUCT #/H
312 20 109 4. 0            P                321               DISCOL BTM PRODUCT CVP
```

Figure 7-11. Multiple-variable tracking (batch) (CRT setup, soft copy).

Description (soft-copy setup). There is some format compatibility between this and the real-time tracking. The soft-copy setup has been shortened to 16 lines of input. The display is used to collect and format the data. Three lines of input have been added to figure the nature of the bursts. A solitary field has been added through which the section request (one of two) is made. For a description of the 16 lines, see the real-time multiple-variable tracking.

The identification of the fields of the three lines includes (each line represents a different report):

Field 1. Descriptive title.

Field 2. (immediate burst). Group or groups to be output (zero no output).

Field 3. (immediate burst). Number of lines of output, zero means to the limit of the pages requested in Field 4 or to the limit of the collected data (last 24 hours).

Field 4. (immediate burst). Maximum number of output pages.

Field 5. (immediate burst). Periodicity of the data and output.

Field 6. (immediate burst). Request to start and stop.

Field 7. (continual bursts). Group or groups to be output (zero no output).

```
REPORT #1  1100  29/02/81
#1 PH CHECKING                    0  0  0    O N      0  0   O N      1  10  1 60    6 Y
#2 DAILY DISCOL OPERATION  0  0  0  O N      2 60 144  Y      0   0   0   0    O N
#3                         0  0  0  O N      0  0   O N      0   0   0   0    O N

555 11    1  6.1   60   20  A       300.0           327.9       360.0  MARES SWEAT ACTUAL #/H
349 11       7.1   60   20  B      3040.0          3053.3      3080.0  WB TO NEUTRALISER #/H
298 11       4.0   25   30  C        575             595         605  FLASH DRUM FWD FLOW CVP
231 11       6.2   25   15  D       10.55           10.96       11.14  NEUTRALIZER PH

567 02          5   20  E      3600.0          4000.8      4400.0  DISCOL FEED #/H
311 20    1  4.0             F                      617              DISCOL FEED CVP
570 02         29   20  G       80.00           94.29       120.00  DISCOL REFLUX GPM
314 20   25  4.0             H                      362              DISCOL REFLUX CVP

569 02         53   20  I      1100.00         1250.73      1400.00  DISCOL TOP PRODUCT #/H
313 20   49  4.0             J                      471              DISCOL TOP PRODUCT CVP
572 02         77   20  K      4500.0          4964.0       5500.0  DISCOL STEAM #/H
316 20   73  4.0             L                      536              DISCOL STEAM CVP

244 20   97  6.1             M                      128.4            DISCOL BTM DEGC
373 20  103  6.2             N                        0.17           ANALYSIS EPR BTM PRODUCT
568 02        113   20  O      3000.0          3210.2       3400.0  DISCOL BTM PRODUCT #/H
312 20  109  4.0             P                      321         =    DISCOL BTM PRODUCT CVP

1000 327.9 3053.3 595 10.96                    D            C           B          A
1001 325.5 3052.3 595 10.96                    D            C         B            A
1002 327.0 3053.0 595 11.03                      D          C         B           A
1003 330.6 3053.1 595 11.06                      D          C         B            A
1004 324.6 3053.2 595 11.07                      D          C          B           A
1005 330.6 3053.6 597 11.09                      D            C         B           A
1006 330.6 3055.2 597 11.07                      D            C         B              A
1007 331.5 3055.0 597 11.03                    D              C             B          A
1008 326.1 3054.6 597 10.96                  D                C             B          A
1009 328.5 3054.2 597  9.72 D                                 C             B          A
1010 322.5 3054.7 597  7.00 D                                 C             B          A
1011 322.5 3054.6 597  7.00 D                                 C             B          A
1012 322.5 3054.3 597  7.00 D                                 C             B          A
1013 324.0 3054.6 597  7.00 D                                 C            B           A
1014 325.5 3054.1 597  7.00 D                                 C            B           A
1015 330.0 3054.0 597  7.00 D                                 C            B            A
1016 335.7 3054.4 597  7.00 D                                 C             B           A
1017 331.2 3054.3 597  7.00 D                                 C             B          A
1018 318.9 3053.8 597 10.68          D                        C             B         A
1019 330.9 3054.6 597 10.70          D                        C             B            A
1020 327.6 3055.0 597  9.62 D                                 C             B           A
1021 328.8 3054.7 597  7.00 D                                 C            B            A
1022 330.3 3054.7 597  7.00 D                                 C            B            A
1023 321.3 3054.7 597  7.00 D                                 C            B           A
1024 320.1 3054.7 597  7.00 D                                 C            B         A
1025 330.6 3055.5 597  7.00 D                                 C             B          A
1026 331.5 3055.5 597  7.00 D                                 C             B          A
1027 328.5 3055.2 597  7.00 D                                 C             B          A
1028 327.9 3053.7 594  7.00 D                        C                   B            A
1029 331.2 3054.6 594  7.00 D                        C                    B           A
1030 325.8 3053.5 594  7.00 D                        C                   B            A
1031 328.2 3053.9 594  7.00 D                        C                   B           A
1032 325.8 3054.4 594  9.76 D                        C                   B           A
1033 329.1 3053.7 594 10.87                        C                   B           A
1034 324.9 3053.1 594 10.86          D               C                 B             A
1035 330.3 3053.6 594  7.00 D                        C                 B            A
1036 322.8 3053.0 594  7.00 D                        C                 B           A
1037 328.2 3053.0 594 10.68       D                  C                 B             A
1038 329.7 3053.6 594 10.94             D            C                 B            A
1039 331.8 3053.4 594 11.01              D           C                 B            A
1040 319.5 3053.5 594 11.05               D          C                 B          A
1041 330.3 3054.1 594 10.99             D            C                 B           A
1042 321.9 3054.3 594 10.96             D            C                 B           A
1043 326.1 3053.3 594 10.97             D            C                 B           A
1044 320.4 3054.0 594 10.86          D               C                 B           A
1045 319.2 3054.3 594  9.53 D                        C                 B           A
1046 316.2 3052.8 594  7.00 D                        C               B            A
1047 319.2 3053.1 594  7.00 D                        C               B            A
1048 319.2 3053.1 594  7.00 D                        C               B            A
1049 333.6 3053.6 594  7.00 D                        C               B           A
1050 320.4 3052.8 594 10.69       D                  C               B          A
1051 327.3 3053.1 594 10.84          D               C               B           A
1052 326.1 3052.5 594 10.94            D             C             B           A
1053 329.4 3053.6 594 11.01              D           C               B           A
1054 326.4 3054.5 594 11.02              D           C             B            A
1055 325.5 3054.1 594 11.03             D            C               B            A
1056 334.8 3054.8 594 10.99           D              C               B           A
1057 328.8 3054.2 594 10.88       D                  C               B          A
1058 332.7 3054.3 594 10.61 D                        C               B          A
1059 324.6 3054.1 594  7.00 D                        C               B            A
```

Figure 7-12. Multiple-variable tracking (batch) (hard copy).

Field   8.   (continual bursts). Number of lines of output (zero means page full).

Field   9.   (continual bursts). Periodicity of the data output.

Field 10.   (continual bursts). Request to start and stop.

Field 11.   (clock burst). Group or groups to be output (zero no output).

Field 12.   (clock burst). Beginning hour.

Field 13.   (clock burst). Frequency in hours apart.

Field 14.   (clock burst). Number of lines of output (zero means page full).

Field 15.   (clock burst). Periodicity of the data output.

Field 16.   (clock burst). Request to start or stop.

(*Hard-copy output*).   The actual formatting of the output will be according to the directions included in the soft-copy setup. Six different reports can be organized simultaneously from the one set of collected values of the 32 chosen variables. Each page of output will be titled with which report it is. It will also include a copy of the setup for future reference purposes.

Use.   Any three combinations of the groups of variables within each section can be output as distinct reports. As the output is neither formatted nor mapped until burst time, the mapping limits, the integer count, and even the number of decimal places can be changed at any time without affecting the stored data. As reports either can be generated from the last 24-hours data at any time or can be generated periodically automatically, the data gathering is very flexible and comprehensive.

Coding.   Each output page must carry the date, time, and a sufficient description of the items tracked. Each row of output should carry also the military time at which it was gathered.

Comments.   The power of mixing both numbers and mapping, particularly of several variables, will become evident to the user when the first control loop is closed. Being able to reconfigure any output in periodicity and constituents will rapidly confirm or eliminate any tentative cause and effect between particular variables.

Only request hard copy when it is needed. The measure of success here is how much hard copy is discarded with only a cursory glance. If this happens frequently, then hard copy is being generated unnecessarily. If *every* hard copy carries pencilled comments or highlights in marker pen, then the output is being utilized fully.

Format 7

Title.   Individual-Variable Investigation (Standard Deviation) (Change Analysis)

Objective.   To collect a sample of size 64 and calculate the mean, the highest value, the lowest value, the range, and the standard deviation. To analyze the data further by studying the changes between the values.

Output example.   Figure 7-13 (soft copy)
Figure 7-14 (hard copy)

Description   (soft-copy setup). The display shows operator-entered directives. It presents both the raw data and the results of the analyses. A row-by-row explanation includes:

Row 1   *Field 1.*   Display title.

Field 2.   Index number of variable to be investigated.

Field 3.   Comment on variable.

Rows 2-9.   An eight-by-eight matrix (filled column by column) of the 64 successive values that are the sample lot.

Row 10   *Field 1.*   Heading.

Field 2.   Actual highest value.

Field 3.   Highest to mean as a percent.

Field 4.   Heading.

Field 5.   Heading.

Field 6.   Lower limit of ordinate for variable value plot.

Field 7.   Upper limit of ordinate for variable value plot.

Row 11.   *Field 1.*   Heading.

Field 2.   Actual lowest value.

Field 3.   Lowest to mean as a percent.

Field 4.   Heading.

Field 5.   Heading.

Field 6.   Actual mean of sample test.

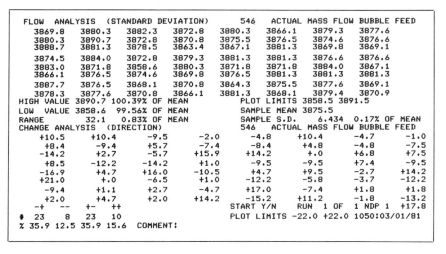

Figure 7-13. Individual-variable investigation (standard deviation and change analysis) (CRT setup, soft copy).

Row 12.   *Field 1.*   Heading.

          Field 2.   Actual range lowest value to highest value.

          Field 3.   Range to mean as a percent.

          Field 4.   Heading.

          Field 5.   Heading.

          Field 6.   Actual standard deviation.

          Field 7.   Standard deviation to mean as a percent.

          Field 8.   Heading.

Row 13.   *Field 1.*   Display title.

          Field 2.   Repeat of the index number of variable under investigation.

          Field 3.   Repeat of comment on variable.

Rows 14-21.   An eight-by-eight matrix (filled column by column) of the 64 successive values of the changes that are the sample lot.

Row 22.   *Field 1.*   Heading.

          Field 2.   Heading.

          Field 3.   Request to initiate sampling run.

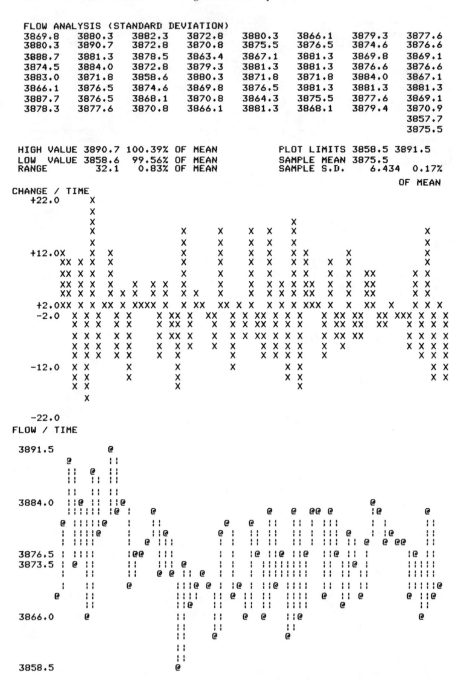

Figure 7-14. Individual-variable investigation (standard deviation and change analysis) (hard copy).

Figure 7-14 continued

Figure 7-14 continued

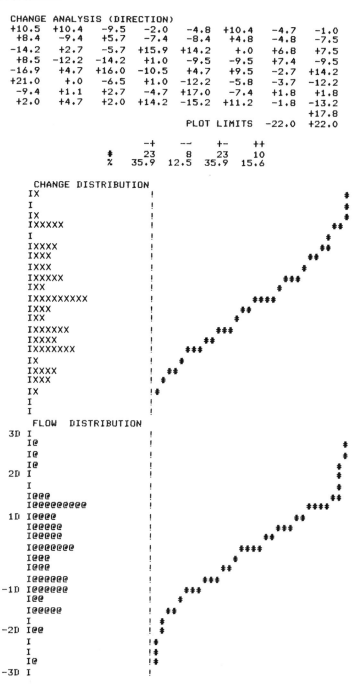

```
CHANGE ANALYSIS (DIRECTION)
+10.5   +10.4    -9.5    -2.0    -4.8   +10.4    -4.7    -1.0
 +8.4    -9.4   +5.7    -7.4    -8.4    +4.8    -4.8    -7.5
-14.2    +2.7    -5.7   +15.9   +14.2    +.0    +6.8    +7.5
 +8.5   -12.2   -14.2    +1.0    -9.5    -9.5    +7.4    -9.5
-16.9    +4.7   +16.0   -10.5    +4.7    +9.5    -2.7   +14.2
+21.0    +.0    -6.5     +1.0   -12.2    -5.8    -3.7   -12.2
 -9.4    +1.1   +2.7     -4.7   +17.0    -7.4    +1.8    +1.8
 +2.0    +4.7   +2.0    +14.2   -15.2   +11.2    -1.8   -13.2
                                                        +17.8
                              PLOT LIMITS    -22.0   +22.0

                    -+      --      +-      ++
                #   23       8      23      10
                %  35.9    12.5    35.9    15.6

        CHANGE DISTRIBUTION
    IX                   !                                 #
    I                    !                                 #
    IX                   !                                 #
    IXXXXX               !                                ##
    I                    !                               #
    IXXXX                !                              ##
    IXXX                 !                            ##
    IXXX                 !                          #
    IXXXXX               !                        ###
    IXX                  !                       #
    IXXXXXXXXX           !                    ####
    IXXX                 !                  ##
    IXX                  !                 #
    IXXXXXX              !              ###
    IXXXX                !             ##
    IXXXXXXX             !           ###
    IX                   !          #
    IXXXX                !       ##
    IXXX                 !      #
    IX                   !#
    I                    !
    I                    !
        FLOW    DISTRIBUTION
 3D I                    !
    I@                   !                                 #
    I@                   !                                 #
    I@                   !                                 #
 2D I                    !                                 #
    I                    !                                 #
    I@@@                 !                                ##
    I@@@@@@@@            !                             ####
 1D I@@@@                !                           ##
    I@@@@@               !                         ###
    I@@@@@               !                        ##
    I@@@@@@@             !                    ####
    I@@@                 !                  #
    I@@@                 !                ##
    I@@@@@@              !             ###
-1D I@@@@@@              !          ###
    I@@                  !         #
    I@@@@@               !       ##
    I                    !  #
-2D I@@                  !  #
    I                    !#
    I                    !#
    I@                   !#
-3D I                    !
```

Field 4/5/6.   Current run number of required number of successive runs.

Field 7/8.   Number of decimal places in the two matrices and the results.

Field 9.   65th value of matrix.

Row 23.   *Field 1*   Heading.

Field 2.   Count of successive changes that go − +(minus plus).

Field 3.   Count of successive changes that go − −(minus minus).

Field 4.   Count of successive changes that go + −(plus minus).

Field 5.   Count of successive changes that go + + (plus plus).

Field 6.   Heading.

Field 7.   Lower limit of ordinate for change value plot.

Field 8.   Upper limit of ordinate for change value plot.

Field 9.   Date and time at beginning of run.

Row 24.   *Field 1.*   Heading.

Field 2/3/4/5.   Percent breakdown, − +, − −, + −, + +.

Field 6.   Heading.

Field 7.   Comment

Use.   To run an analysis, request the display. Enter the index number of the variable to be investigated, an annotating comment, the number of successive runs to be made, and the request to start. At the conclusion of the 66th cycle, the mean, standard deviation, and "change counts" will be calculated and appear within the display. If the analysis is one of several runs, a complete hard-copy report prints and the next run begins. If it is an individual run, hard copy is by request.

Coding.   If a run has been requested, 66 successive values of the variable are collected. The mean and standard deviation are calculated from the first 64 samples, as are the highest value, the lowest value, and the range. For display purposes, the percents are calculated. For the change analysis, the differences between the 66 successive values are calculated. The changes have direction as well as magnitude. The 64 pairs of two ordered changes that occur within the 64 sets of three successive values that are the 66 member sample group are counted into whichever of the four combinations they belong $(-+, --, +-, ++)$. For display purposes, the breakdown of percents is calculated.

Description (hard copy). The upper section and lower section of the soft-copy setup appear side by side in the upper third of the hard copy. The middle third of the hard copy consists of the changes plotted against time and their distribution by magnitude and direction. The lower third of the hard copy is a plot of the sample values against time and their distribution. The ranges of the ordinates of both sets of plots are entered on the soft-copy setup.

Use (hard copy). An archival-quality record of the sample and the results for future reference and comparison.

Coding (hard copy). For either of the time plots, the ordinate values are calculated from the entered values of the range limits. As each of the 20 intervals in each time plot is printed, the sample group is tested and, for any sample found within the interval, an asterisk is printed in the column representing the time interval. The number of samples falling in that interval are counted and presented as a horizontal bar at the end of the line. The headings and coordinates are formatted as necessary during the printing of the output.

Comment. Only when flows and variables are subjected to rigorous analysis can the improvements of any proposed control scheme be quantified and verified. Because of the difference in the nature of the effects of short-term disturbances and long-term disturbances, actual improvements in the short term (without analysis) can be obscured by the longer-term disturbances. Also, without analysis, control schemes on occasions may be incorrectly blamed for overall plant performance degradation that is in fact caused by unrelated events. Finally, the question of improvement is no longer subjective. The hard facts, untinged by wishful thinking, can be assembled that in themselves prove or disprove.

Comment on Results of "CHANGE" Analysis

The purpose of the exercise was to forecast the next flow, given the two preceding flows. The direction and magnitude of the forthcoming change is predicted from the direction and magnitude of the current change. As can be seen, the distribution of the magnitude and direction of the changes is flat and symmetrical. Because of the flatness of the distribution, a change of any certain magnitude is just as likely as a change of any other magnitude—whatever the magnitude of the preceding change. Simply stated: magnitude prediction is not possible.

But there was a surprise in the results. For any change in a given direction, a change in the opposite direction was twice as likely as a continuation in the same direction. This was unexpected and further data was collected. The findings were surprisingly consistent run to run. The similarity of results for different flows was astonishing. Any bias caused by the periodicity of the sampling was ruled out after a doubling of the time interval gave identical results. The implications of these results are studied further in the next chapter. The emphasis to be made here is that no assumptions, however basic, can be made at any time until real data from the process has been collected through, and processed by, the computer. That data then must be studied carefully by the people implementing the process computer control before any control schemes can be given trial runs.

Format 8

Title. Multiple-Variable Investigation

Objective. To sample several variables simultaneously and repeatedly and analyze for standard deviation, mean, and changes.

Output example. See Figure 7-15.

Description (*soft-copy setup—identical to hard-copy output*). There are 16 units in the display. Each unit is identical in format. For each unit:

Line 1, Field 1. Index number of variable to be investigated.
Line 1, Field 2. Number of cycles between collecting individual samples.
Line 1, Field 3. Military time for sample collection to begin.
Line 2, Field 1. Mean of sample.
Line 2, Field 2. Standard deviation of sample.
Line 3. Subheading.
Line 4, Fields 1/2/3/4. Change analysis of sample in percents.
Line 5. Operator enterable comment.

Use. By the operator entering index numbers, times, periodicities, and comments, the 16 units can be programmed to run as required. It is possible to run all 16 simultaneously on 16 different variables, or to run the 16 one

```
******************** ******************** ******************** *******************
546 6 0850           546 6 0934           545 6 0850           545 6 0934
3978.5      10.7     3966.5      12.9     699.8       1.32     693.5       0.95
-+    --    +-   ++  -+    --    +-   ++  -+    --    +-   ++  -+    --    +-   ++
32.8 14.1 32.8 20.3 34.4 21.9 34.4  9.4 32.8  9.4 32.8 25.0 34.4 12.5 34.4 18.8

******************** ******************** ******************** *******************
546 6 0901           546 6 0945           545 6 0901           545 6 0945
3859.5      11.4     3980.9      16.5     694.8       2.41     698.6       2.29
-+    --    +-   ++  -+    --    +-   ++  -+    --    +-   ++  -+    --    +-   ++
32.8 14.1 32.8 20.3 34.4 17.2 35.9 12.5 23.4 25.0 23.4 28.1 25.0 32.8 25.0 17.2
                     ANALOG BACKUP                            MASSIVE UPSET
******************** ******************** ******************** *******************
546 6 0912           546 6 0956           545 6 0912           545 6 0956
3875.5       6.4     3969.5      10.8     697.7       2.32     697.5       1.88
-+    --    +-   ++  -+    --    +-   ++  -+    --    +-   ++  -+    --    +-   ++
35.9 12.5 35.9 15.6 31.3 14.1 31.3 23.4 29.7 15.6 31.3 23.4 23.4 25.0 25.0 26.6
MANUAL                                                        RECOVERING
******************** ******************** ******************** *******************
546 6 0923           546 6 1007           545 6 0923           545 6 1007
3852.6       9.9     3990.4      12.7     693.8       1.92     696.4       1.38
-+    --    +-   ++  -+    --    +-   ++  -+    --    +-   ++  -+    --    +-   ++
31.3 20.3 32.8 15.6 34.4 14.1 32.8 18.8 31.3 15.6 31.3 21.9 28.1 21.9 28.1 21.9
                                                             ALMOST LINED OUT
```

Figure 7-15. Multiple-variable investigation (soft copy).

after the other on just one variable, or to run only four simultaneously on four different variables and repeat three times. The hard copy is made by request.

Coding. The act of entering the index number of the variable clears the particular unit of earlier data, results, and comment. At the entered time, the sampling will begin at the entered periodicity. At sampling's end the mean, standard deviation, and change analysis are calculated and become available for display.

Comments. It is up to the ingenuity of an individual user to get the best benefit from this type of program. If there were three similar flows (11, 12, 13) and three alternative strategies (A, B, and C), then a test could be scheduled: at 1000 hours—A on 11, B on 12, C on 13; at 1030 hours—B on 11, C on 12, A on 13; and at 1100 hours—C on 11, A on 12, B on 13. The results may confirm that one strategy is superior to the others on each of the three similar flows. The results may confirm that one flow is better than the other two whatever strategy is used. There may be nothing conclusive. That is part of the fascination. At times the results can be quite unexpected. Remember, with programmable processor power, it is easy to run a second series.

Example #1

(All numbers will be carried to seven significant places.)
An algorithm for calculating the density (RHO) of ethylene from the gas'
Rankine temperature and absolute pressure is:

Let temperature in Rankine degrees $= T$, and absolute pressure in pounds
per square inch $= P$

1         $D = P/(T*10.73350)$
          $E1 = T*10.73350$
          $E2 = 0.8919800*10.73350*T - 12593.60 - 1602280000./(T*T)$

          $E3 = 2.206780*10.73350*T - 15645.50$
          $E4 = 15645.50*0.7316610$
          $E5 = 4133600000./(T*T)$

2         $E6 = D*D*D*E4$
          $E7 = 2.368400*D*D$
     $EPWR = EXP(-E7)$
   $ERROR1 = D*(E1+D*(E2+D*(E3+E6+E5*EPWR*(1+E7))))$

          If$/(ERROR1/P)-1/ < 0.0005$ GO TO 3

   $ERROR2 = E1+D*(2*E2+D*(3*E3+6*E6+E5*EPWR*(3+E7*(3-2*E7))))$
        $D = D-(ERROR1-P)/ERROR2$ GO TO 2

3    $RHO = 28.04900*10.73350*D/10.73147$
For $T = 500$, $P = 120$:
          $D =$ 0.02235990
          $E1 =$ 5366.750
          $E2 =$ -14215.69
          $E3 =$ -3802.263
          $E4 =$ 11447.20
          $E5 =$ 16534.40

FIRST ITERATION:
ERROR1:
          $E6 =$ 0.1279702
          $E7 =$ 0.001184117
       $EPWR =$ 0.998817
   $D*D*D*E5 =$ 0.1848409
$EPWR*(1+E7) =$ 0.9999997

$ERROR1 =$ $D*E1$ 120.0000
             $+D*D*E2$ -7.107349
             $+D*D*D*E3$ -0.04250615
             $+D*D*D*E6$ 0.000001430600
             $+D*D*D*E5*EPWR* (1+E7)$ 0.1848407

```
        ERROR1 =                              113.0350
(ERROR1 /P)−1                                −0.05804167
```

```
NOTE:               E1 =   5366.750
  D*D*E5*EPWR*(1+E7) =      8.266621   FULL CALCULATION
  D*D*E5             =      8.266623   ABBREVIATED METHOD
```

```
ERROR2:
        D*D*E5 =           8.266623
        EPWR*E7 =          0.001182716
        EPWR*E7*E7 =       0.000001400474
```

```
        ERROR2 =    E1                     5366.750
                  +2*D*E2                 −635.7224
                  +3*D*D*E3               −5.702997
                  +6*D*D*E6                0.0003838838
                  +3*D*D*E5*EPWR          24.77053
                  +3*D*D*E5*EPWR*E7        0.02933121
                  −2*D*D*E5*EPWR*E7*E7    −0.00002315439
```

```
        ERROR2 =                           475.0125
            D = D                           0.02235990
                −(ERROR1−P)/ERROR2          0.001466277
            D =                             0.02089362
```

```
NOTE:                        E1 =  5366.750
D*D*E5*EPWR*(3+3*E7−2*E7*E7)=    24.79984 FULL CALCULATION

D*D*E5*3                =        24.79987 ABBREVIATED
                                          CALCULATION
```

:FIRST ITERATION CONCLUDED

On close inspection of ERROR1,

```
        D*D*D*D*D*E4<<<E1
D*D*E5*EXP(−2.368400*D*D) * (1+2.368400*D*D)
                = D*D*E5*1*(1+0)
                = D*D*E5
                    in significance with E1
```

Further, on close inspection of ERROR2,

```
        6*D*D*D*D*D*E4<<<E1
D*D*E5*EXP(−2.368400*D*D)*(3+2.368400*D*D* (3−2*2.368400*D*D))
                = D*D*E5*1*(3+0)
                = D*D*E5*3
                    in significance with E1
```

The terms dropped in the previous expressions actually affect digits in the eighth and greater significant place of the final answer. Therefore, ERROR1 and ERROR2 can be rewritten:

ERROR1 = D*(E1+D*(E2+D*(E3+E5)))
ERROR2 = E1+D*(2*E2+D*3*(E3+E5))

The abbreviated expressions give a shorter program that can be executed more quickly and contains no library calls.

RHO was calculated by the full algorithm and by the abbreviated algorithm for each element of a six-temperature-by-eight-pressure matrix. The particular Fahrenheit temperatures used were 0, 40, 80, 120, 160, and 200. The pressures (psig) used were 90, 120, 150, 180, 210, 240, 270, and 300. A portion of the matrix is presented in Table 7-8. In the table the results are presented to seven significant figures. At no location of the original matrix was there a difference between the RHOs. The conclusion can be drawn that the original algorithm's accuracy was beyond the capabilities of 23-bit floating math. The implementation of the full original algorithm was a needless sophistication, which wasted machine resources.

Table 7-8
The Density of Ethylene for Various Temperatures and Pressures

	Method	90 psig	120 psig	150 psig
0 degF	FULL	0.6409678	0.8450639	1.060126
	ABB	0.6409678	0.8450639	1.060126
40 degF	FULL	0.5786974	0.7573213	0.9431719
	ABB	0.5786974	0.7573213	0.9431719
80 degF	FULL	0.5289574	0.6892942	0.8539547
	ABB	0.5289574	0.6892942	0.8539547

Example #2

A 23-bit floating math cannot add 32,767,540 and 1.331 together. If the addition were attempted, the result would be 32,767,540. The 1.331 is beyond the seven significant digits in 32,767,540. This is not as great a drawback as it may appear. After all, the greater number is 24 million times as great as the smaller; a considerable difference in orders of magnitude. Can it be a problem in integrating process flows by computer? Not really. This example is worked to show that the choice of time interval can and does have a more important effect on flow integration results than the loss of significance by addition of numbers of dissimilar orders of magnitude.

Assume that in a 24-hour period a flow was held steady for the first 22 hours at a rate of 198,600 lb/day, and for the last two hours it flowed at 43,323 lb/day (see Case 1, Case 2, and Case 3).

For each of the cases, the average flow is the same. The loss of some digits in the summing does not materially affect the answer. But, as in all sampling procedures, infrequent sampling can give misleading results. Let us assume that the flow was stopped from 20.15.06 to 20.46.54, a 31-minute, 48-second stoppage during the first 22 hours (see Case 4, Case 5, and Case 6).

CASE 1	Average of 24 counts		One hour apart
	Counts		Sum
First 22 hours	22		4369200
Last 2 hours	2		86646
Total 24 hours	24		4455946
		Average	185,660 lb/day

CASE 2	Average of 1440 counts		One minute apart
	Counts		Sum
First 22 hours	1320		262152000
Last one hr, 59 min	119		5155437
Final one minute	1		43323
Total 24 hours	1440		267350700
		Average	185,660 lb

CASE 3	Average of 17280 counts		Five seconds apart
	Counts		Sum
First 22 hours	15840		3145824000
Last One hr, 59 min, 55 seconds	1439		62341797
Final five seconds	1		43323
Total 24 hours	17280		3208208000
		Average	185,660 lb

Case 4	Average of 24 counts		One hour apart
	Counts		**Sum**
First 22 hours			4369200
(including stoppage)	22 of 198600		
Last two hours	2 of 43323		86646
Total 24 hours	24		4454846
No stoppage seen		Average	185,660 lb

Case 5	Average of 1440 counts		One minute apart
	Counts		**Sum**
First 22 hours			255975400
(excluding stoppage)	1289 of 198600		
Stoppage	31 of 0		0
Last two hours	120 of 43323		5198750
Total 24 hours	1440		261174100
A stoppage of 31 minutes		Average	181,371 lb

Case 6	Average of 17280 counts		Five seconds apart
	Counts		**Sum**
First 22 hours			
(excluding stoppage)	15459 of 198600		3070157000
Stoppage	381 of 0		0
Last 2 hours	1440 of 43323		62385110
Total 24 hours	17280		3132542000
A stoppage of 31 minutes,			
45 seconds		Average	81,281 lb

The result in Case 5 is not materially different from the result in Case 6. The missing of one in an overall count of 1440 is an error of 0.07%. However, in Case 4, if a stoppage of five seconds had occurred at the precise moment of an hour count, the overall error would have been 4%. That is a relatively large error. If the period of the count were reduced to eight minutes, the error of one missed count will be 0.5% of the daily figure—not an unreasonable error, but still one that can be made smaller.

For integration of daily flows, one-minute intervals are a satisfactory compromise between processor overhead and accuracy.

These equations with adjustable factors can be implemented on the computer too. But to work with *e* is to ignore the power of the computer. Pneumatic analog controllers must, by the nature of their construction (and if they are to serve in universal applications), work with *e*, the difference between the actual value and the desired value. Electronic analog controllers (until recently) have merely duplicated the pneumatic mechanism with electronics. Programmable processors do not have to work with *e* alone. In fact, they are already cognizant of the set point (the desired value) and the actual value (measurement) as two separate and distinct entities. They are forced to do so by the nature of the entries of the values into the machine. The latter is an analog input, the former is a keyboard entry.

One other flaw in analog controllers must be noted. The generation of *e* does not distinguish between changes in the desired value and changes in the actual value. Consequently, if the desired value is changed suddenly, the system will be disturbed seriously—a control action was predicated, *e* had changed suddenly. Any changes required in the desired value must be taken slowly if the system is to be disturbed the least.

For any given flow system that includes a control valve, a pump, and a piping system, the flow within the system will come to a stable equilibrium about some mean flow for any one position of the valve, provided the pressure differential across the ends of a piping system does not vary. Any valve movement at all will perturb this equilibrium and cause greater fluctuations to occur. Without valve action, it is a self-correcting equilibrium with natural dampening. If the flow exceeds the equilibrium flow, the pump head is reduced, and the flow friction is increased. Both act to reduce the flow. If the flow is below the equilibrium flow, the pump head is increased, and the flow friction is reduced. Both act in concert to restore the equilibrium flow.

Analog controllers cannot take preemptive corrective action; they can respond only to error conditions. Therefore, a flow with the control valve set at a fixed position on manual will fluctuate less than a flow with a control valve under automatic controller control. The standard deviations of the two flows will corroborate this point. Readers will acknowledge that chemical plant operators make use of this phenomenon more often than plant management would wish. To control a flow ideally, the valve would have to act in concert (but in an opposite sense) with the fluctuating pressure differential across the ends of the piping system, thereby maintaining a constant net differential—responding to the changing condition that caused the flow fluctuations (acting in a primary sense to the causal), rather than responding to the resulting flow fluctuations (acting in a secondary sense to the caused).

If the corrective-action algorithm based on Equation (8-2) were implemented for fluid flow (on a computer), and the factors changed to provide data for a regression analysis, the analysis would provide optimum values.

These optimum values would be: $K_2 = 0$; K_1 equals such a value that based on the conditions in existence at $t = 0$, the corrective action to the valve would place the valve as quickly as possible at the position considered necessary for the actual flow to be coincident with the desired flow: a full 100% immediate corrective-control action. This assumes that Δt is a longer period than the time it takes the flow system to come to a new equilibrium.

Given that a fluid flow is to be controlled, it is necessary to know four things for it to be controlled satisfactorily. First, the actual fluid flow must be known. This will be called the ACTUALFLOW. Second, the target fluid flow must be known. Call this TARGETFLOW. Third, the position of the valve for the current actual flow must be known. Ordinarily, the air pressure going to the valve is used as an indirect indication of the valve position. Call this ACTUALCVP. Finally, there must be some inferred position of the valve (or actuator air pressure) that will result in the TARGETFLOW. Call it CALCULATEDCVP. The difference between ACTUALFLOW and the TARGETFLOW is the ERROR. The difference between the CALCULATEDCVP and the ACTUALCVP is the CHANGE.

That is, in analog control the value of ERROR is used as a predictor for some value of CHANGE such that when that value of CHANGE is entered into the system, the resulting ACTUALFLOW will be closer to the TARGETFLOW. In properly implemented digital control (which means discrete time intervals apart) the value of CHANGE can be calculated to make ACTUALFLOW coincident with TARGETFLOW. Obviously, digital logic has an enhanced capability of timing as well as of sizing the desired changes. Analog control does not. The timing of CHANGE or incremental parts thereof can be made independent of the timing of ERROR.

The deficiencies of automatic analog controllers can be summarized as follows:

1. They can only respond to error conditions. No preemptive action is possible since the measured variable must deviate to generate correction.

2. The control action is symmetrical about the desired value.

3. The control action is linear about the desired value.

4. The control action is predicated on the differential pressure measurement's being a true indicator of the real flow. This is not true in three respects: (a) the signal is proportional to the square of the velocity of the flow, (b) that proportionality is *not* constant over the total range of that signal, and (c) the signal is also a function of the flowing density.

5. Any given tuning factors for the controller are effective for a very limited range of the measurement value only.

6. A choice of "no-action" at any time is not possible.

7. The speed and the magnitude of the correction are insolubly tied to the speed and the magnitude of the error.

8. The controllers cannot distinguish between changes in the desired value (the set point) and changes in the actual value; they react only to the difference.

9. They maintain a constant differential pressure across the orifice plate which is neither constant volume flow nor constant mass flow. Realizing this is very important, they do not maintain a true, constant volume flow—a constant chart flow is not a constant volume flow unless the density of the flow is constant.

10. The controller permits only one set point, at one discrete point.

11. The dead-time of the process cannot be accommodated. The current error is used to generate the immediate correction. The results of the correction will not be measured as an error until some time later. This is definitely a problem with temperature control and any form of concentration control. Analog control of a process with dead-time results in a sinusoidal fluctuation pattern of both error and correction.

12. Any modifications to a control scheme will require physical removal and installation of wires, tubes, and instruments. It will require minutes or hours, and some instruments may be taken out of service for a short period.

13. The controller can neither monitor its own performance nor accommodate its own deficiencies.

14. It is virtually impossible to orchestrate more than two controllers to act in unity of purpose.

15. Macrocontrol is unfeasible.

The power of programmable processor control stems from:

1. Each control loop can have its own customized decision tree that predicates "action" or "no-action" for any given set of circumstances and further describes what such action is to be.

2. Preemptive action is possible. Causal events may be modified rather than caused events corrected. Timely correction can prevent measured variable fluctuations.

3. The control action need not be linear, symmetrical, or about one unique desired value (set point).

4. Control action *can* result in constant volume flow or constant mass flow if enough is known about the orifice plate's coefficient of discharge and the flowing fluid density.

5. Control action can be written that is correct and efficient at any time for any part of the range of the measurement value.

6. The speed and the magnitude of the correction is absolutely divorced from the speed and magnitude of the error.

7. The movements of the measured variable are distinguished from changes in the desired value. Large changes in the desired value can be entered into the system without disruption.

8. The dead-time of the process is easily accommodated. Any control action and its resulting error are easily correlated.

9. Any modifications to a control scheme may require only keyboard entries. It is done in seconds or minutes—nothing need be taken out of service.

10. The processor can monitor its own performance and can make accommodation for the deficiencies.

11. Any number of controllers can be made to act in a concerted fashion. True process-unit control is possible.

12. Macrocontrol is implemented easily.

Figure 7-14 presents evidence that the short-term flow as seen by the computer is neither sinusoidal nor monotonic in direction or increments. The flow is staggered, a reverse movement being twice as likely as a continued movement. The magnitude of the increment of change, within limits, is random. Further, movement of the valve one pulse-increment (even on a 1000-increment, full-stroke scale) CAN and DOES result in significant process changes. The flow itself has the least short-term dispersion if there is no control valve movement. "Short-term" is understood to mean that the series of flow measurements has no discernible drift. "Long-term" is understood to mean that if the series of flow measurements has a discernible drift, the drift itself has a pattern that is repeated enough so that no overall effect is manifest.

Careful perusal of the figures in Chapter 7 will provide further ample evidence of the nature of the flow measurements and the sensitivity of the process.

Without knowledge of the chemical process or the loop to be controlled, it is possible to hypothesize a "time-velocity" algorithm for process control by a programmable processor. It is not a recommended approach. It has both logical and practical deficiencies which will be indicated. It is described because it exists and it is used.

Time-Velocity Algorithm (not recommended)

For a pseudo-digitized analog controller,

$$\text{DELTA} = \text{PROPAF}* ((\text{TARGET} - \text{ACTUAL}) \quad\quad (8\text{-}3)$$
$$- \text{RESTAF} *(\text{ACTUAL} - \text{HISTORY}))$$

$$\text{ACTION} = \frac{\text{DELTA}}{\text{ACTUAL}} * \text{REGP} \quad\quad (8\text{-}4)$$

where:

TARGET = required value of the controlled variable

ACTUAL = current measured value of the controlled variable

HISTORY = measured value of the controlled variable last cycle

PROPAF = proportional action factor

RESTAF = reset action factor

DELTA = required change in the measured value of this cycle

ACTION = number of increments of change to the regulator this cycle

REGP = actual position of the regulator in increments

The second term of Equation (8-3) produces a corrective action proportional to the error. The third term lessens or heightens this action, depending on whether the actual value is moving to or away from the target value. Equation (8-4) relates DELTA to the current position in the range of possible values.

This algorithm is unsatisfactory. The reset action term actually disrupts smooth control. In measurements of the system at regular discrete time intervals the term (ACTUAL-HISTORY) is without pattern or predictability. If the measured value cycles faster than the measurement period, this term definitely will misrepresent actuality. If the required change (DELTA) is not set equal to the current error (TARGET-ACTUAL), the full meanings of the measurements are not being used. The system will play catch up forever. There are further problems with the nature of the corrective action. While a deviation exists, a succession of sets of correction increments (ACTION) will be sent to the regulator. If the elapsed time the system takes to correct is many times the measurement and calculation period, serious overcontrol will ensue. A correction of one increment sent 10 successive times will be a nine-increment overcorrection. No fractional increment corrections are possible. Corrections can be made only in whole numbers of increments. If the action factors (PROPAF and RESTAF) are chosen deliberately for their smallness (to prevent excessive action), in fact they will result only in action when the error is great enough to have the fractional increment of correction rounded to one. Such a correction of small magnitude will be less than sufficient. The overall result of small factors is a wide dead-band and totally insufficient correction to either side of it.

The conversion process of the measured value's required change to corrective increments does not reflect that although the regulator scale is linear, the measured variable will be porportional to flow—a square-root scale. The ACTUAL is proportional to the square root of REGP. This is true whether we are considering a valve or a set point. Equation (8-4) should be rewritten

$$\text{ACTION} = 2 * \frac{\text{DELTA}}{\text{ACTUAL}} * \text{REGP}$$

from

$$y = kx^2, \quad \frac{dy}{dx} = 2kx = \frac{2kx^2}{x} = \frac{2y}{x}, \quad dy = 2y\frac{dx}{x}$$

Finally, the algorithm itself may be misused. Because of the deficiencies already described, it certainly should not be used in direct positioning of the valve, that is, direct digital control (DDC). Should it be used in positioning the set point, that is, supervisory set-point control (SSC), great care must be taken. The controller with a supervised set point is already attempting to accommodate and correct flow changes. Do not compensate for flow deviations through the set point. The set point needs adjusting only when the flowing conditions alter (pressure, temperature, purity), and when an actual change in the TARGET occurs. Changes in the controlled value in themselves are strictly the analog controller's duty. Under supervisory set-point control, the TARGET is the required value of the set point, and the ACTUAL is the current measured value of the *set point*. This can, does, and will result in offsets between what is wanted and what results.

The question remains: Should flowing condition changes and target value changes be made instantaneously or ramped? Flowing condition changes almost always should be made immediately. The changes are measured to exist. Any sluggish response will be the fault of the analog controller, and it is the analog controller that must be dealt with to correct the problem. Target value changes should be ramped slowly and deliberately. Frequent changes back and forth to the set point of an analog controller are to be avoided. They wear out the mechanisms and increase the width of an already wide dead-band.

A better approach to formulating an algorithm follows:

Dead-Reckoning Trajectorial Algorithm

This algorithm is

$$TREG = f(TARGET)$$

$$ACTION = TREG - REGP$$

and f can be solved from the current conditions by

$$REGP = f(ACTUAL)$$

where:

$TREG$ = target position in increments for the regulator
that will result in the required value
of the controlled variable
$REGP$ = actual position in increments of the regulator
$TARGET$ = required value of the controlled variable
$ACTUAL$ = current measured value of the
controlled variable
$ACTION$ = number of increments of change to
the regulator this cycle

This is a superior algorithm. No offset is possible. It controls to that position of the regulator which will give the closet possible value of the controlled variable to the required value of the variable. It uses the best available knowledge—the current conditions—and also can be adapted to accommodate process-lag or dead time.

The function $f(TARGET)$ should be modified only when the results are judged. Therefore, in the dead-time period between executing ACTION and evaluating the resultant output the function $f(TARGET)$ must remain unaltered. The measured output cannot be used to alter the input at the moment of the measurement. The trajectory must be set and the range checked before the trajectory can be adjusted in a meaningful fashion. This does not preclude $f(TARGET)$ being a predictive algorithm for adjusting the regulator according to changing inputs on a feed-forward basis. However, the exact form of the algorithm will be altered on a discontinuous, lag-adjusted, feedback basis. Using an output measurement to adjust the input of the moment introduces deliberate and possibly unnecessary cycling.

Formulating the DRT Function

This function will comprise two math models, a compensatory feedback math model and a predictive feed-forward math model (see Figure 8-2). The latter model will give a result that would be ideal if all the derivation assumptions in the model were correct, if the model incorporated all the knowledge necessary for a true computation, and if nature resembled the math model equivalent. An ideal result is unlikely. Therefore, the former model is used to correlate the real results with the known inputs and to make compensations in the future predictions. Obviously, as knowledge of any one control loop is gained, the predictive math model can be made increasingly sophisticated, and less reliance will be placed on the feedback model.

The function need not confine itself to one discrete target value. It may be advantageous to have a target range between two set points. The algorithm may have several parts, each part constituting a different control action for different circumstances. The control action may be asymmetrical, slow-acting, and linear on one side of the target range; fast-acting and non-linear on the other. The algorithm may ignore all minor, short-duration disturbances and correct only for long-term drift.

The algorithm can be made sophisticated by including counters for duration and delay measurements. Only when some measured variable is outside a limit for a certain number of consecutive cycles will a control action be instituted. Or, a certain control action has been executed and for a certain delay period no further action will be taken. This allows a correction to pass through the system (dead-time) or allows an equilibrium for the new conditions to be reached.

Whatever the final form of the algorithm, its overall accuracy will depend on fluid flow measurement and fluid flow control. With a full knowledge of the flows and practicable flow control, the laws of the conservation of mass and the conservation of energy will permit effective level control and temperature control, respectively.

It simplifies matters greatly if it is considered that all control loops involve regulating fluid flow. Let the fluid be called the *direct* variable. In those loops where some variable other than flow is of primary interest let it be called the *indirect* variable. Therefore, in a "flow-control" loop the flow itself is the direct variable, and there is no indirect variable. In a "temperature control" loop the temperature is the indirect variable, and the flow of the heating or

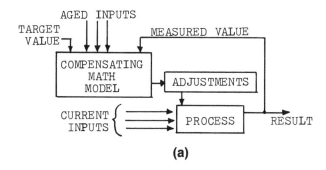

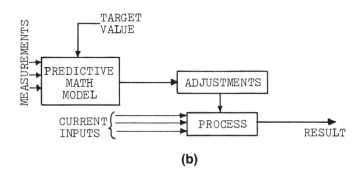

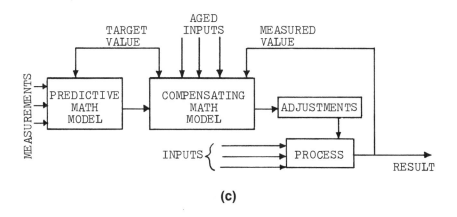

Figure 8-2. (a) A schematic of feed-back control
(b) A schematic of feed-forward control
(c) A schematic of combined feed-forward and feed-back control

cooling medium the direct variable. In a "level-control" loop the level is the indirect variable, and the flow to or from the vessel is the direct variable. In a "pH-control" loop the pH is the indirect variable, and the regulated flow is the direct variable.

The four options in implementing control strategy on a programmable processor are.

1. *Supervisory set-point control*—time-velocity algorithm
2. *Supervisory set-point control*—trajectorial algorithm
3. *Direct digital control*—time-velocity algorithm
4. *Direct digital control*—trajectorial algorithm

The time-velocity algorithm is not recommended in either case. Figure 8-4 is an example of a trajectorial algorithm for supervisor set-point control. Figure 8-5 is an example of a trajectorial algorithm for direct digital control. Both algorithms involve decision trees (see Figure 8-3). Pneumatic analog controllers under supervisory set-point control are insensitive to miniscule adjustments and take long periods of time to adjust to small changes. Electronic controllers do not have these problems.

Flow, Control Valves, and Flow Control

A relationship does exist between valve position (air pressure to the valve is actually measured) and the resulting fluid flow in a liquid/piping system. Chapter 2 made some preliminary observations. This relationship will now be explored in depth.

It is assumed that any control valve to be used will be sized and adjusted correctly. This means that it will take a definite signal to commence opening the valve, there will be no hysteresis and, because of the overall effect of the valve-plus-piping-system, equal increments in valve position will result in decreasing increments of flow. It also means that the piping system is capable of a maximum flow well in excess of the maximum flow possible with the installed control valve wide open.

Figure 8-6 shows such a relationship. Triangles *OAB* and *oab* emphasize that at different positions in the relationship, the incremental increase of flow rate per incremental increase of valve position varies considerably. If a linear relationship between the flow rate and the valve position is assumed, then for any particular actual flow rate (ACTFR) and actual valve position (ACTVP)

SIMPLE TEST 1 current value 1 history value	COMPOUND TEST 1 current value 2 history values		DECISION

SIMPLE COMPARISONS:
(no dead band)

	RISING	RISING	ACTION #1
RISING	RISING	FALLING	ACTION #2
FALLING	FALLING	RISING	ACTION #3
	FALLING	FALLING	ACTION #4

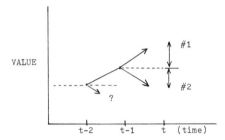

SOPHISTICATED COMPARISONS:
(no dead band)

	RISING FASTER	ACTION I
	RISING SLOWER	ACTION II
RISING	FALLING SLOWER THAN IT ROSE	ACTION III
	FALLING FASTER THAN IT ROSE	ACTION IV
	RISING FASTER THAN IT FELL	ACTION V
	RISING SLOWER THAN IT FELL	ACTION VI
FALLING	FALLING SLOWER	ACTION VII
	FALLING FASTER	ACTION VIII

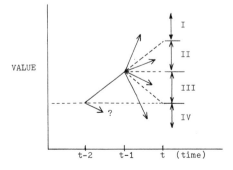

Figure 8-3. The compounding of simple comparisons can result in sophisticated decision trees, particularly with the inclusion of dead-band.   *Figure 8-3 continued*

Figure 8-3 continued

```
SOPHISTICATED COMPARISONS:
 (with dead band)
```

```
                               RISING FASTER                  ACTION A
                               RISING SLOWER                  ACTION B
                  RISING       STEADY                         ACTION C
                               FALLING SLOWER THAN IT ROSE    ACTION D
                               FALLING FASTER THAN IT ROSE    ACTION E

                               RISING                         ACTION F
                  STEADY       STEADY                         ACTION G
                               FALLING                        ACTION H

                               RISING FASTER THAN IT FELL     ACTION I
                               RISING SLOWER THAN IT FELL     ACTION J
                  FALLING      STEADY                         ACTION K
                               FALLING SLOWER                 ACTION L
                               FALLING FASTER                 ACTION M
```

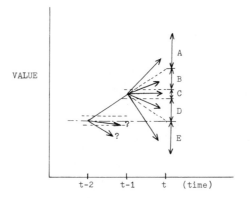

the valve position for any target flow rate (TGTFR) can be calculated (CALCVP)

$$CALCVP = ACTVP * \frac{TGTFR}{ACTFR}$$

where:

CALCVP = calculated valve position
 ACTVP = actual valve position
 TGTFR = target flow rate
 ACTFR = actual flow rate

or

$$OB' = OB * \frac{A'B'}{AB} \text{ (See Figure 8-7a, 8-7b, 8-7c, and 8-7d)} \qquad (8-5)$$

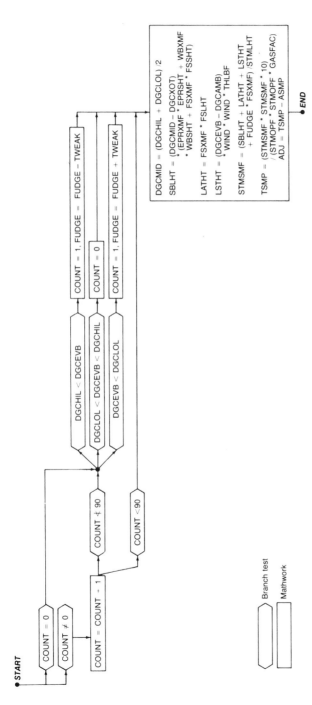

Figure 8-4. Temperature control by a set-point supervisory control algorithm.

Figure 8-4

1. This is strictly a control algorithm for closed-loop supervisory control of the evaporator bottoms' temperature in the Witches' Brew Process. Mass flows are calculated previous to the algorithm. The daily material balances and alarming are not part of this algorithm.

2. The logic diagram does not show the extra logic required to track the process when this loop is off the computer. Tracking logic is required to provide bumpless transfer when the loop is switched to closed-loop control. Set-point supervisory control will not provide "soft" manual, and it cannot actively control flow in the strict sense. It only can set the set point of a regular analog controller.

3. Control Summary: A heat balance is struck around the evaporator using the term with FUDGE to accommodate any discrepancy. The use of the counter allows a prior adjustment of FUDGE to work through the system before a subsequent adjustment is made. The heat requirements of the evaporator load set the steam feed rate, and the back calculation gives the necessary steam set-point position. As knowledge of the physical data and the process grow, better predictions of the heat requirements become possible, and the need for the term containing FUDGE will shrink.

4. A controller set point has two coincident scales: the square root 0-10 chart divisions of flow scale, and the linear 0-1000 stepping motor position scale.

The equivalency is (set-point chart divisions)^2*10 = set-point stepping motor position from Equation (2-8):

set-point flow = $OPF * \sqrt{GASFAC} * DIVV$

Rearranging:

$$DIVV = \frac{\text{setpoint flow}}{OPF * \sqrt{GASFAC}}$$

Combining:

$$\text{stepping motor position} = \frac{\text{set-point flow * set-point flow * 10}}{OPF * OPF * GASFAC}$$

COUNT   Interadjustment delay counter (counts up to 15 minutes), manipulated in algorithm.

DGCEVB Temperature of evaporator bottoms, analog input measurement.

DGCHIL Evaporator bottoms high-temperature limit, set through CRT at 91degC.

DGCLOL Evaporator bottoms low-temperature limit, set through CRT at 89degC.

FUDGE   An algorithm-adjusted factor used to fine tune the heat balance across the evaporator.

TWEAK   An arbitrary quantity used periodically to adjust FUDGE incrementally when the evaporator is out of temperature limits, an initial input.

DGCMID Average of temperature limits, calculated.

SBLHT   Heat rate to raise temperature of feed, calculated.

LATHT   Heat rate to evaporate Frogs' Spit, calculated.

LSTHT   Heat lost to environment (an experimental correlation of windspeed and temperature undergoing trial), calculated.

DGCXOT Temperature of extract, analog input measurement.

EPRXMF EPR mass flow in extract, calculated from measurements.

EPRSHT EPR liquid specific heat Btu/lb/· degC, an initial input.

WBXMF   Witches' Brew mass flow in extract, calculated from measurements.

WBSHT   Witches' Brew liquid specific heat Btu/lb/· degC, an initial input.

FSXMF   Frogs' Spit mass flow in extract, calculated from measurements.

FSSHT   Frogs' Spit liquid specific heat Btu/lb/· degC, an initial input.

Figure 8-4 continued

Figure 8-4 continued

FSLHT Frogs' Spit latent heat of vaporization Btu/lb, an initial input.

DGCAMB Ambient air temperature, analog input measurement.

WIND Wind velocity, analog input measurement.

THLBF Tower heat loss constant of proportionality, an initial input.

STMSMF The mass flow of steam at the required set point, derived in algorithm.

STMLHT The latent heat of condensation of reboiler steam Btu/lb, an initial input.

TSMP Target set-point stepping motor position (kliks), derived in algorithm.

STMOPF Reboiler steam orifice plate factor, an initial input.

GASFAC Correction factor for flowing steam conditions, calculated.

ASMP Actual set-point stepping motor position (kliks), analog input measurement.

ADJ Steam set-point adjustment in stepping motor kliks, result of algorithm.

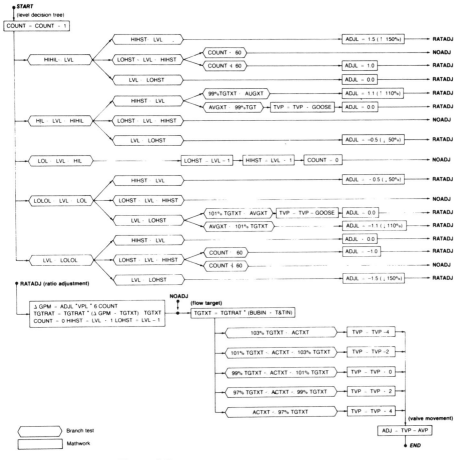

Figure 8-5. Level control by a DDC algorithm.

Figure 8-5

1. This is strictly the control logic for closed-loop control of the Witches' Brew Process reactor level. All the mass flows are calculated previous to the algorithm. The daily material balances and alarming are not part of this algorithm.

2. This logic diagram does not show the extra logic required to track the process when this loop is off the computer, nor does it feature any extra logic to provide anything beyond the level reset flow ratio closed-loop control. Tracking logic is required to provide bumpless transfer when the loop is switched to closed loop control.

3. Control Summary: the gain or loss of level is used to reset the target mass flow ratio according to a many-branched decision tree. The target ratio is used to predict the target flow out of the reactor, given the sum of the actual flows in. Any divergence between the target flow and the actual flow is corrected by valve ramps of various rates.

4. The design of the algorithm is to maintain as smooth a flow out of the reactor as possible. Consequently, level fluctuations in the wide middle deadband are ignored, partially catered to in an asymmetrical fashion in the adjacent "slo" bands, and definitely reacted to at the extremes.

ACTLVL Actual level in %, analog input measurement.
BUBIN Flow rate in gpm of Bubble to reactor, calculated from measurements.
T&TIN Flow rate in gpm of Toil and Trouble to reactor, calculated from measurements.
ACTXT Actual flow rate out of reactor in gpm, calculated from measurements.
AVGXT Recent average flow rate out of reactor in gpm, calculated from the six most recent actuals.
TGTXT Target flow rate out of reactor in gpm, derived in algorithm.
AVP Actual valve position kliks (0 close to 1000 open), analog input measurement.
TVP Target valve position kliks (0 close to 1000 open), derived in algorithm.
TGTRAT Ratio of target flow rate out to actual flow in, derived in algorithm.
HIHIL High high level limit, operator set through CRT at 75%.
HIL High level limit, operator set through CRT at 60%.
LOL Low level limit, operator set through CRT at 40%.
LOLOL Low low level limit, operator set through CRT at 25%.
LOHST Level history low ⎰ ⎱ set at actual level ± 1% in that cycle that the cumulative
HIHST Level history high ⎱ ⎰ drift of the level triggered an adjustment to the ratio.

VPL Reactor volume in gallons per 1% level, an initial input.
GOOSE Arbitrary valve position adjustment of about three kliks, used to nudge system, an initial input sized by trial and error.
ADJ Adjustment in kliks sent to control valve, result of algorithm.
COUNT Count used to calculate rate at which level is changing (6 counts = 1 minute), zero when ratio adjusted.

For the actual flow rate (FR_A) and actual valve position (VP_A), a calculated valve position (VP_C) for the target flow rate (FR_T) results in the flow rate (FR_R). This is satisfactory in Figures 8-7a and 8-7b. The result is closer to, but has not overshot, the target. It is not satisfactory in Figures 8-7c and 8-7d. The resulting flow rate (FR_R) overshoots the target, although it is closer to the target flow than was the initial flow.

An algorithm based on repeated use of Equation (8-5) would result eventually in the achievement of the target flow. For high-flow rates, the target

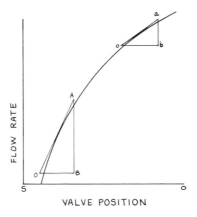

VALVE POSITION

Figure 8-6. The relationship of flow rate and valve position.

would be approached directly in decreasing steps (Figure 8-8a). For low-flow rates, the target would be overshot in decreasing steps (Figure 8-8b) in almost all instances. It can happen that the overshooting will not decrease, and the target is circled repeatedly (Figure 8-8c).

If the linear relationship between flow rate and control valve position was assumed to originate from that valve position where flow begins ($0'$), then results similar to those described for Figures 8-7a and 8-7b still obtain and are true for the whole range. The unsatisfactory results described in Figures 8-7c and 8-7d are not possible. Figures 8-9a and 8-9b show the linear relationship originating from $0'$, the valve position where flow begins, for initial flows above and below the target flow.

If a control algorithm were based on this offset linear relationship, then great care would be needed to establish the exact position of $0'$. Exactly established, the algorithm is robust and able to satisfy wide divergence between initial flow amd target flow (Figure 8-10).

Inaccurately established to the left of the true $0'$, it can cause a severe overshoot problem for a low initial flow rate and a high-target rate (Figure 8-11). Inaccurately established to the right of $0'$, the algorithm will not work at low-flow rates; it will give smaller, less accurate movements at medium-flow rates and barely be equal to the algorithm with a correct $0'$ at high-flow rates (Figure 8-12).

The best approach is a composite algorithm (Figure 8-13). At low-flow rates, control is strictly by rote. At medium-flow rates, control is based on Equation (8-5) on origin $0'$, with fractional proration to prevent overshoot. At high-flow rates, control is based on Equation (8-5) on origin $0'$, without proration. As shown, the algorithm is extremely accurate for small errors and quite impervious to large errors. It is strictly a flow algorithm.

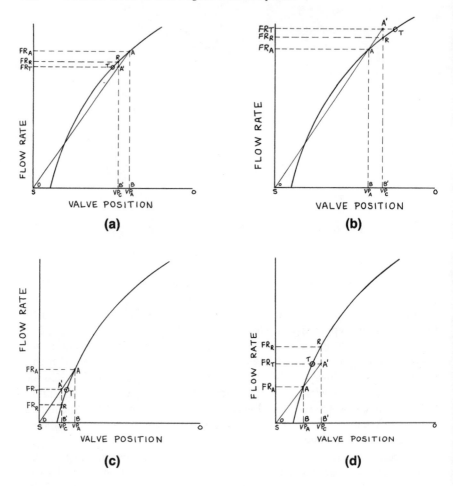

Figure 8-7. (a) a geometric construction of flow rates and valve positions for actual, target, and result: given a prorated linear correction with the initial conditions of actual flow being *above* the target and both in the *upper* ranges of valve position and flow rate.

(b) a geometric construction of flow rates and valve positions for actual, target, and result: given a prorated linear correction with the initial conditions of actual flow being *below* the target and both in the *upper* ranges of valve position and flow rate.

(c) a geometric construction of flow rates and valve positions for actual, target, and result: given a prorated linear correction with the initial conditions of actual flow being *above* the target and both in the *lower* ranges of valve position and flow rate.

(d) a geometric construction of flow rates and valve positions for actual, target, and result: given a prorated linear correction with the initial conditions of actual flow being *below* the target and both in the *lower* ranges of valve position and flow rate.

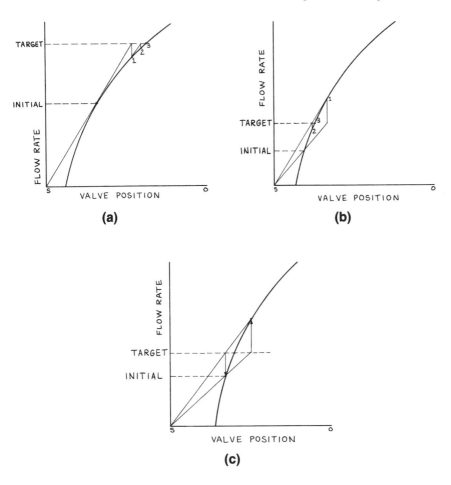

Figure 8-8. (a) a geometric construction showing the actual flow converging on the target without overshoot as a result of a series of corrections, each based on a linear proration.

(b) a geometric construction showing the actual flow converging on the target with decreasing overshoot as a result of a series of corrections, each based on a linear proration.

(c) a geometric construction showing the unusual circumstance of the actual flow circling the target: corrections on either side based on linear prorations, each subsequent correction returning the system to the alternate position.

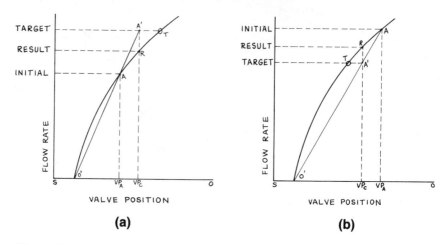

Figure 8-9. (a) a geometric construction showing flow rates and valve positions for actual, target, and result: given a prorated linear correction based on a false origin, with the initial conditions of actual flow being *below* the target.

(b) a geometric construction showing flow rates and valve positions for actual, target, and result: given a prorated linear correction based on a false origin, with the initial conditions of actual flow being *above* the target.

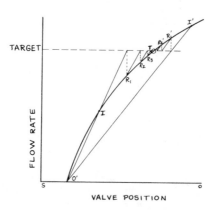

Figure 8-10. A geometric construction showing (whether the actual flow is initially above or below the target) the actual flow converging on the target without overshoot as a series of corrections, each based on a linear proration with a precisely set false origin.

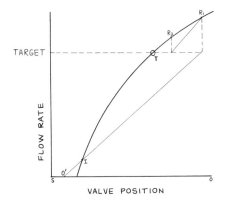

Figure 8-11. A geometric construction highlighting the initial overshoot possible with both an underset false origin and a gross discrepancy between the initial and target flows. However, there is a subsequent convergence.

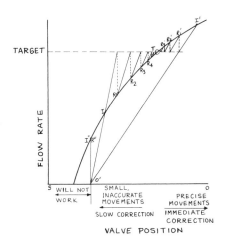

Figure 8-12. A geometric construction showing (whether the actual flow is initially above or below the target) the actual flow converging on the target without overshoot as a result of a series of corrections, each based on a linear proration with an overset false origin, *provided* the initial conditions are beyond the false origin.

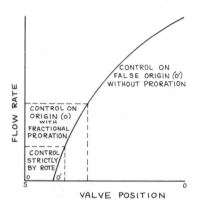

Figure 8-13. The zoning of alternative logic for processor-based control algorithms.

Project: Close One Loop

Make a very careful assessment of the problem. What is it? Where is it? Why is it a problem? Make written notes. Successful computer control involves money and effort. These are scarce resources and should be spent only on the loops where the returns are the greatest. Write an objective for the project. What can be achieved by putting this loop on the computer? Is it safety, dollars, capacity, productivity, or quality?

Make a sketch of the process unit and its instrumentation. List all the field-measured variables you think you need to know to solve the problem. List all the calculated variables you need to know. Revise these two lists in terms of what is already in the computer. Generate a list of what new instrumentation is needed, its cost, and its installation costs. Is any extra hardware required for the computer? Do the proposed benefits exceed the anticipated costs? If not, do not waste any more time.

Write, assemble, debug, and load the programs necessary to facilitate tying in the new instrumentation and generating the new calculated variables. Now begin studying the process and collecting data through the computer. Use data output program #1 (single-input real-time checking) to be certain that all input variables are correctly calculated and that their inputs are properly calibrated. Use the real-time data output programs #3 and #4 to monitor the variations and patterns of perturbation of all the variables under study. Can cause and effect relationships be established? The plots make identifying lag-times and event times very easy. Also, the plots can expand even narrow ranges of interest to expose the most minor of fluctuations—fluctuations that are not visible on traditional recorder charts. The output of actual numbers exposes the actual order of magnitude in all the changes.

Use of the batch data output programs #5 and #6 allows long-term data to be collected. They also enable unattended data logging around the clock and over weekends and holidays. Only when enough observations have been made should the control algorithm be formulated. Is there any further instrumentation or computer hardware necessary to close the loop? Does it affect the economics of the project? Continue only if the benefits are shown still to be greater than the costs.

Make provisions for the display of the loop's parameters and variables on a screen. It is through this display that supervision of the control loop will be exercised. It will be necessary to display:

1. The control loop's name
2. The actual value of the program-calculated or field-measured "indirect" variable that is ultimately controlled
3. The target value or acceptable limits for the "indirect" variable
4. The mode of the loop
5. The condition of the loop
6. The actual value of the program-calculated or field-measured "direct" variable that is manipulated directly
7. The target value or acceptable limits for the "direct" variable
8. The actual flow under control if not already displayed
9. The control action in increments sent
10. The position of the regulator

The following should be accessible but not necessarily on the same display:

1. The target position of the regulator
2. The value of any ratios used in coordinating one flow with another
3. The flow divisions
4. The orifice plate factor for the flow
5. All the counts and counters associated with this flow
6. Any factors in the algorithm that may require changing

Assemble and load the display. Install all necessary computer hardware and instrumentation. Write, assemble, debug, and load the control algorithm.

Now actual computer supervision of the control loop can begin. Monitor the performance of the algorithm with any of the data output programs, particularly program #2, triple-variable real-time tracking. In this program make one parameter the regulator position, the second parameter the actual value of the direct variable, and the third parameter the actual value of the indirect variable. Modify and restructure the control algorithm until the desired performance is achieved. Use the variable investigation programs (#7 and #8) to provide hard data on which form of the algorithm actually gives the best performance. Stop when the original objectives are achieved.

It is a possibility that during the investigative process, simple non-computer changes in either instrumentation or the process suggest themselves. If they are implemented, the objective of the project may be attained without computer involvement. Stop the project at once and move to the next loop.

Traditional control maintains constant chart flow readings, and the process-end variable, even if it is meant to hold steady, inevitably will wander. Computer control with its feed-forward and feedback components can hold that process-end variable steady. But if it does hold it steady, then all the incoming, previously unseen fluctuations will have been corrected. Inevitably, this means that the flow readings on the charts will *not* remain steady. Effective computer control results in steady outputs and unsteady inputs. This is very disconcerting to people unfamiliar with the outward manifestations of computer control. There are other features that are disconcerting, particularly where gas flows are concerned.

Gas flow measurements are supposed to be more accurate if the downstream conditions at the orifice plate are measured. In fact, a change in the formula of the flow calculation makes the upstream conditions just as acceptable. But reliance on mass flows calculated from the downstream pressure cause overaction. If the pressure drops, the flow is increased and the differential pressure drop increases. This in turn drops the downstream pressure measurement even further, which increases the flow, and so on. More stable control occurs if orifice plate upstream conditions are measured. But even so, headers that branch cause problems. Greater or lesser flow across any tee branch will cause the assumed conditions in the branch to vary. Individual process trains that are parallel may be found to have operating characteristics that are a function of the production rates of the other trains.

If the computer ratio controls a flow to a flow that is controlled by an analog controller, the former will have much larger than expected fluctuations. Analog controllers need an error condition before they can correct. Computer control is precise. Consequently, the minor fluctuations of the one will be magnified in the other.

Once a control algorithm is proven, it may be replicated on other towers or reactors if for the new loops the equipment is identical. Without doubt, each piece of equipment has its own personality, and if the equipment is not identical (and if the instrumentation in any way varies), the algorithm will have to be retailored for better performance. There is no such thing as a typical distillation column or a typical reactor. Therefore, there cannot be a typical control system that is applicable to all columns or all reactors. But the control of one loop at a steady production rate is but one element in an overall scheme of process control.

For any individual flow within a chemical process, three different conditions can be recognized: lined-out operation, the orderly change between two different rates of lined-out operation, and the disequilibrium of startup/shutdown. In terms that are more precise, eight different states can be defined:

1. Fully lined-out operation, no adjustment necessary
2. Almost lined-out operation, minor upward trim adjustment required
3. Almost lined-out operation, minor downward trim adjustment required
4. A major change of rate upward required
5. A major change of rate downward required
6. Startup (from no flow to that first rate of flow that is stable)
7. Shutdown (from that last rate of flow that is stable to no flow)
8. No flow

For a process as a whole that will contain two or more flows, six different states can be defined:

1. Lined-out operation at a stated production rate
2. The ramping upward of a process to a higher stated production rate
3. The ramping downward of a process to a lower stated production rate
4. The startup of a process from a zero production rate to the lowest stable rate of production
5. The shutdown of a process from the lowest stable rate of production to a zero production rate
6. A zero production rate

Obviously, the process states come about indirectly. It is only by direct manipulation of individual flows that changes in the process states occur. Total computer control of any one flow would mean directing and shepherding that flow from no flow to startup to some stable flow, upward and downward through various rates of stable flow, to shutdown and an eventual return to no flow. Total computer control of a process would mean taking a process from zero production rate through process startup onward to a stable rate of production, herding the process through several, if not many, stated rates of production to eventual process shutdown and a return to a zero production rate. The maintenance of a single flow at lined-out operation is but a minor part of the complex overall task of total process computer control. But before computer control is explored in detail, an appreciation of some other composite systems will be beneficial.

Two composite systems in which the actions of an individual are severely curtailed by the presence of others are a convoy of ships and a car on a crowded freeway. Whatever the ability to turn, or the safest top speed of an individual, the actions of the total group are governed by the ship with the widest turning circle or the car with the lowest speed. Any attention given to the worst performer to improve his record raises the performance of the group—a convoy is no faster than its slowest ship. Likewise, any actions taken by any individual other than the worst performer must be taken with the realization that anything but short-term performance ultimately will be governed by the worst performer. So it is in concerted process control.

A unit or plant cannot have its overall production rate ramped more than at that rate which is comfortable for the flow with the greatest problems. To put it another way, a computer system can control the overall process only as well as it can contain the detrimental side effects from the loop with the worst performance.

In every complete control action either material or energy has its rate of transfer changed, which changes the conditions of the sending system as well as the conditions of the receiving system. Consequently, in balancing one part of the plant some other part must become unbalanced. Paraphrased, the second law of thermodynamics states that order comes from greater disorder elsewhere. This backward and forward shuffling between units can be seen at times on the freeways. In process control it could mean that to ignore minor imbalances is the best control strategy because to attempt to correct them can only throw some other systems out of balance. This in turn will affect their supporting systems. The disturbances spread further and further until the consequences are sufficiently diminished as to be imperceptible.

With composite systems, the greatest advantages are achieved not by tuning one "superloop," but by making modest improvements on the loops with the poorest performance.

Macrocontrol

The ultimate in computer control is total plant control—macrocontrol. This can be attempted only on the secure foundation of precise, predictable microcontrol (control on each of the individual loops). Obviously, the total program package would be able to start the process from zero production and return it to zero production, but also it must be able to handle equipment, instrument, and even computer hardware failure and malfunction. Distributed microprocessor control gives much greater process security against these problems than does one giant plant computer. Each microprocessor would handle a unit operation or section of the plant. The overall coordination of the several microprocessors would be by one master microprocessor.

It would be an unusual set of circumstances if any individual economic maximum or minimum in the plant did not have a broad peak. This is one reason why there is no economic penalty with large tolerances between set points on any of the control loops. If the peak is broad, then there is little point in optimizing a unit or process every one to five minutes. The continual adjusting of production rates may in fact carry a cost penalty. It does take energy and effort (and therefore money) to change rates. It also cycles and fatigues the process equipment. It is well within the bounds of possibility that *one microprocessor* could supervise and optimize 250 microprocessor-controlled unit operations. An individual optimization each shift or once a day would be sufficient.

Macrocontrol would lead to the very long-term improvement of the plant. It would be necessary to run statistical correlations on product quality, raw material yields, and process conditions. Because there is a long history of each process condition stored in the computer, this actually becomes a simple task. Any required optimum reflux ratios, solute solvent ratios, column temperatures and pressures, reactor conditions, or excess tolerances can be calculated from a CRT in the plant office. The calculation of any one of those parameters by manual methods would be virtually an impossible task.

All control loops generate their greatest savings in the initial stages. It is a case of diminishing returns. The best approach to automating a plant is not to get all the savings possible out of one loop and then pass on to the next, but to get half the savings on each of the loops before squeezing any more savings out of the first loop. Then repeat the cycle as often as necessary to achieve the original objectives. Always look at the marginal returns. How much *extra* benefit does the *extra* effort produce? Put that extra effort wherever the extra benefits are the greatest. Some projects may be mutually exclusive, other projects may complement one another perfectly—so plan the effort with the future in mind. Implement control strategy and develop computer expertise to get the most effective return on the effort. Do not let intellectual satisfaction or curiosity wreak havoc with your schedule and your budget. Exercise dollar control on all proposed projects. Computers and programmers are scarce resources. Do not waste them.

Witches' Brew Computer Control
For all the control loops that will be described in Figures 8-14, 8-15, 8-16, and 8-17, the control valves themselves are acted upon directly by the computer. By definition, this is DDC or direct digital control. These loops are not set-point supervisory control. There are no active analog controllers between the computer and the control valves. In the Witches' Brew plant, conventional

set-point analog controllers are physically installed between the computer and the control valve. However, under a fall-back scheme, they become active when, and only when, the computer — for whatever reason — is taken off-line.

Each of these figures represents a snapshot in time of the process. For each single processor cycle, the inputs are measured, the flows calculated, the decisions taken, the adjustments sent out, and the displays generated. Consequently, for any single display, the data is particular to one single processor cycle in time, and as such the data is a matched set. EVERYTHING described within that display occurred in just ONE processor cycle. The measurements, the calculations, the decisions, the adjustments, the displays — all will be repeated in their entirety in the very next processor cycle. The display describes only one cycle; do not read it as a several-cycle average synopsis; do not read it as a concatenation of time-elapsed events — there are no missing sequences. If the processor cycle time is 10 seconds, then in 24 hours each of these displays will be generated afresh 8640 times. For each single time, the explanation accompanying each of these figures would have to be rewritten. The adjustments shown in an immediately previous display become incorporated into the inputs measured and calculated in the display that is its immediate successor in time. Material in this section uses the phrase, "this measurement calculation decision cycle." By definition it is incomplete, although all measurements, calculations, adjustments, and displays are complete. However, the cycle is still in progress, and no fresher information is available. Do not confuse it with the previous, complete cycle.

CONTROL:	LOOP CONDITIONS					REACTION		1050	29 FEB 81	
	MODE	ADJ	GPM	#/h	CVP	TARGET	TVP	VARIABLE		DIV
DOUBLE	C	-1		625.1	273	622.7	272	6.022:1	RATIO	8.13
BUBBLE	C		13.60	3750.2	380		380	3750	LOAD	7.55
TL&TBL	C	3	9.20	625.3	699	628.0	702	1.009:1	RATIO	7.24
RXRCYC	L		150.88		511	150.88	511			7.53
RX-LVL	L		25.51		367	25.51	367	77.7	LEVEL%	3.94
RX-STM	M			2610.4	350	2610.4	350	124.2	DEGC	6.83
FLSCND	M		160		496	160	496	49.3	DEGC	7.08
FLSCLR	M		125		***	125	999	48.6	DEGC	4.75
ACIDCP	A	1	21.21		236	22.22	237	61.1	LEVEL%	3.36
MSWEAT	W		4.70	375.5	931	380.7	931	7.2	PH	5.74

Figure 8-14. A display of the control loop conditions at the reaction end.

This display is a concise record of the current control activities of the 10 loops on the computer at the reaction end of the process. All the important control mode, flow, and adjustment information on each of the loops is immediately available. For each loop, the mode and the basic flow measurement are presented where they can easily be found — at the ends of the line. Any adjustment currently being made is the second item in from the left on a line. It is well spaced so it stands out. The second item (a pair) in from the right is the information necessary to close the loop. It is a ratio-for-control or the current value of a feedback measurement. On the left-hand side of the display, the third and fourth items in are the calculated actual flows, volume and mass. Almost dead center on each line is the target flow with a current valve position on the left and a target valve position on the right. In the following descriptions the sequence of the written presentation is not always identical section to section, and it does not correspond to the sequence of items on a row. It follows the logic in the algorithm behind the control loop. It was felt strict conformity between loops and by the row order would at times confuse more than it clarified.

Double is on computer feed-forward ratio control with no significant problems (MODE C). An adjustment of one pulse close (ADJ − 1) was sent to the Double control valve in this measurement calculation decision cycle. The actual flow measured is 8.13 divisions of gas flow (DIV 8.13). For the current flowing conditions of temperature, pressure, and purity, this calculates to be 625.1 pounds per hour of Double (#/h 625.1). The current control valve position is 273 pulses open (CVP 273). The Double target flow of 622.7 pounds per hour (TARGET 622.7) is back-calculated from the Bubble target (3750 LOAD) and the set Bubble:Double ratio (6.022:1 RATIO). From the actual flow, the target flow, and the current valve position, the target valve position of 272 pulses open is calculated (TVP 272). It is the pulse difference between the current valve position and the target valve position that is issued as an adjustment. The set Bubble:Double ratio of 6.022 was entered by the operator. That value was found to be optimal after a study of reactor conditions and reactor yields was made. If the process requirements were to alter, then this ratio could be changed simply by moving the cursor on the video display to that field and entering the new value through the keyboard. It will then remain at that entered value until it is altered to another operator-entered value at some time in the future.

Bubble is on computer flow control with no significant problems (MODE C). It was not found necessary to adjust the control valve position of 380 pulses open (CVP 380), since the control valve position was "on-target." Hence, no adjustments are shown (it is blank under ADJ). The actual flow measured is 7.55 divisions of liquid flow (DIV 7.55). For the current Bubble density and strength, this calculates to be 13.60 gallons per minute solution volume flow (GPM 13.60) and 3750.2 pounds per hour Bubble mass flow (#/h 3750.2). At the reaction end of the Witches' Brew process, Bubble is the key flow, so the optimum reactor strategy requires that Double be ratioed to Bubble and, in turn, Toil and Trouble be ratioed to Double. Consequently, setting the rate at which Bubble is fed to the reactor automatically sets the rates for Double and Toil and Trouble. The production plans for the plant (the plant loading) require that Bubble be fed at 3750 pounds per hour (3750 LOAD). As with the Bubble:Double ratio, the Bubble load rate is operator entered by cursor positioning and keyboard entry. Changes in Bubble load are accommodated by a ramp described in Figure 8-15. As the actual mass flow is within the control-algorithm-prescribed tolerance limits for the mass flow target, the target valve position of 380 (TVP 380) must itself be "dead on target." Because Bubble is not undergoing a ramp, no target mass flow is shown (blank under TARGET), it is understood in such circumstances that the LOAD is simultaneously the intermediate and final mass flow target.

The Toil and Trouble mixture (TL&TBL) is on computer feed-forward ratio control with no significant problems (MODE C). An adjustment of three pulses open (ADJ 3) was sent to the Toil & Trouble control valve in this measurement calculation decision cycle. The actual flow measured is 7.24 divisions of liquid flow (DIV 7.24). For the current mixture density, this calculates to be 9.20 gallons per minute volume flow (GPM 9.20). For the current Toil purity, it calculates to a Toil mass flow of 625.3 pounds per hour (#/h 625.3). The current valve position is 699 pulses open (CVP 699). The Toil target flow of 628.0 pounds per hour (TARGET 628.0) is calculated from the Double target (TARGET 622.7) and the set Toil:Double ratio (1.009:1 RATIO). From the actual flow, the target flow, and the current valve position, the target valve position of 702 pulses open is calculated (TVP 702). It is the pulse difference between the current valve position and the target valve position that is issued as an adjustment. The set ratio of 1.009 was entered by the operator and will remain at that value until altered at some time in the future. That value was found to be optimal after a study of reactor conditions and reactor yields. The actual chosen value is operator entered by cursor positioning and keyboard entry.

The reactor recycle (RXRCYC) is not on computer control. It is on local (MODE L), that is, under the control of the panelboard-mounted conventional analog controller. The actual flow measured is 7.53 divisions of liquid flow (DIV 7.53). For the current reactor conditions, this represents an actual volume flow of 150.88 gallons per minute (GPM 150.88). The current control valve position is 511 pulses open (CVP 511). To provide "bumpless-transfer" should this control loop be switched to the computer in this processor cycle, the computer monitors the actual volume flow and the current valve position and sets the target flow and the target valve position to their respective current actual values (TARGET 150.88) (TVP 511). There are no computer-generated adjustments and so none are shown (ADJ blank).

The reactor level (RX-LVL) is not on computer control. It is on local (MODE L), that is, under the control of the panelboard-mounted conventional analog controller. The actual reactor forward flow measured is 3.94 divisions of liquid flow (DIV 3.94). For the current reactor conditions, this represents an actual volume flow of 25.51 gallons per minute (GPM 25.51). The current control valve position is 367 pulses open (CVP 367). The actual reactor level is at 77.7% (77.7 LEVEL%). To provide "bumpless-transfer" should this control loop be switched to the computer in this processor cycle, the computer monitors actual volume flow and the current valve position and sets the target flow and the target valve position to their respective current actual values (TARGET 25.51) (TVP 367). There are no computer-generated adjustments and so none are shown (ADJ blank).

The reactor steam (RX-STM) is on maintain valve position computer control (MODE M). In this case the valve is moved so that its position coincides with the target valve position. The operator can position the valve by entering the desired target valve position (TVP) with cursor positioning and keyboard entry or by striking dedicated keys that offer up, down, fast up, fast down options. Striking such keys causes the computer itself to add or remove certain numbers of pulse positions from the target valve position and then to move the valve accordingly. No adjustments are shown (ADJ blank) because the current valve position (CVP 350) is "on-target" for the target valve position (TVP 350). The actual flow measured is 6.83 divisions of flow (DIV 6.83). For the current flowing conditions of steam temperature and pressure, the steam mass flow calculates to 2610.4 pounds per hour (#/h 2610.4). To provide "bumpless-transfer" should the mode be changed in this processor cycle, the target flow is set equal to the actual mass flow (TARGET 2610.4). The mode itself is changed by the operator with cursor positioning and keyboard entry or by striking dedicated mode keys. Striking a mode key causes the computer to change the mode of the loop to the mode required/struck. The actual temperature at the bottom of the reactor is 124.2 degrees Centigrade (124.2 DEGC).

The flash drum condenser cooling water flow (FLSCND) is on maintain valve position computer control (MODE M). No adjustments are shown (ADJ blank) because the current valve position of 496 pulses open (CVP 496) is "on target" for the target valve position (TVP 496). The actual flow measured is 7.08 divisions of flow (DIV 7.08). This calculates to a volume flow of 160 gallons per minute (GPM 160). To provide "bumpless-transfer" should the mode be changed in this processor cycle, the target flow is set equal to the actual volume flow (TARGET 160). The actual temperature of the vapors leaving the condenser is 49.3 degrees Centigrade (49.3 DEGC).

The flash drum cooler cooling water flow (FLSCLR) is on maintain valve position computer control (MODE M). No adjustments are shown (ADJ blank) because the current valve position is wide open (CVP ***). Because it is possible to put less than 3 psig and more than 15 psig on a valve positioner and because it is possible to underdrive and overdrive stepping motors, the following arbitrary designations are made: Given it is a 0 to 1000-pulse range, 0 is considered to be a closed, underdriven valve with less than 3 psig on it; 1000 is considered to be a wide-open, overdriven valve with more than 15 psig on it; 1 is considered to be a just-closed valve with 3 psig exactly on it; and 999 is considered to be a just-wide-open valve with 15 psig exactly on it. The subtle but important difference is that a single up-pulse sent to a valve at a pulse position of 1 will move it to a pulse position of 2, while a single up-pulse sent to a valve at a pulse position of 0 may or may not move it to a pulse position of 1. It may only reduce how much the valve or stepping motor is underdriven. At the other end of the scale a single down-pulse sent to a valve at a pulse position of 999 will move it to a pulse position of 998, while a single down-pulse sent to a valve at a pulse position of 1000 may or may not move it to a pulse position of 999. It may just reduce how much the valve or stepping motor is overdriven. The flash drum cooler cooling water flow control valve is wide open at 999 and is flagged as being so by ***. The valve position is also "on target" because the target valve position is 999 (TVP 999). The actual flow measured is 4.75

divisions of flow (DIV 4.75). This calculates to a volume flow of 125 gallons per minute (GPM 125). To provide "bumpless-transfer" should the mode be changed in this processor cycle, the target flow is set equal to the actual volume flow (TARGET 125). The actual temperature of the liquid at the bottom of the flash drum is 48.6 degrees Centigrade (48.6 DEGC).

The acidic crude product flow from the flash drum to the mixer (ACIDCP) is on closed loop feedback control but is above its upper limit (MODE A). An adjustment of one pulse up (ADJ 1) was sent to the acidic crude product control valve in this measurement calculation decision cycle. The actual flow measured is 3.36 divisions of liquid flow (DIV 3.36). For the current flowing density, this rate calculates to be 21.21 gallons per minute (GPM 21.21). The current control valve position is 236 pulses open (CVP 236). The acidic crude product target flow is 22.22 gallons per minute (TARGET 22.22). This target flow rate was decided many cycles ago by the control algorithm as being sufficient to reduce the then high level in the flash drum within a short time; however, it was not such a dramatic change in flowrate as to cause problems elsewhere in the process. From the actual flow, the target flow, and the current valve position, the target valve position of 237 (TVP 237) was calculated as necessary if the actual flow was to be brought "on target." It is the pulse difference between the current valve position and the target valve position that is issued as an adjustment.

Mares' Sweat (MSWEAT) is on closed loop feedback control in a wait period (MODE W). In an earlier cycle a radical change was made in the Mares' Sweat target flow, and a dead-time counter was set into effect. No further changes to the target flow can be made until this counter has run out. When the counter runs out, the consequences of the change should be evident and measured by the feedback signal. No adjustments are made (ADJ blank). The actual flow measured is 5.74 divisions of flow (DIV 5.74). For the currrent Mares's Sweat density, this rate calculates to volume flow of 4.70 gallons per minute (GPM 4.70) and a mass flow of 375.5 pounds per hour (#/h 375.5). The current control valve position is 931 pulses open (CVP 931). The Mares' Sweat target flow as set at the beginning of the dead-time count several cycles previous is 380.7 pounds per hour (TARGET 380.7). Because the pH was high and is currently between limits, but the current actual mass flow is below target, the logic in the algorithm will leave the target valve position equal to 931, its previous cycle value (TVP 931). The current valve position is "on target," and no adjustment is made. This type of logic is built into the control algorithm for the "W" mode. The pH of the mixer effluent is currently 7.2 (7.2 pH).

The variable values within each of the 10 control algorithms for the reaction end of the process are given in the display of Figure 8-15. The row titles and relative row positions corrrespond to Figure 8-14. However, the Figure 8-15 display will be explained column by column and not row by row. The majority of values in this display are operator alterable by cursor positioning and keyboard entry.

```
CONTROL:  FACTORS & CLOCKS              REACTION          1052    29 FEB 81

          LO LIMIT HI         LTR   RAMP   ESC     OOL    WAIT  MIN&MAX   OPF

DOUBLE                        6.000              .  .    .  .   .  .               76.89
BUBBLE                               102    1 . 2  .  .   .  .                      1.80
TL&TBL                        1.000              .  .    .  .   .  .                1.27

RXRCYC          150                                3    .  .   .  .               20.04
RX-LVL    20.0   80.0  H 77.7                   .  .    .  .   .  .                6.47
RX-STM   120.0  122.0                             10    .  .   .  .              382.20

FLSCND    48.0   52.0                           .  .    .  .   .  .    40<F       22.60
FLSCLR    48.0   52.0                           .  .    .  .   .  .    60<F       26.32

ACIDCP    40.0   60.0  H 65.4                    .  .    .  .   .  .                6.31
MSWEAT     7.0    7.4  H  7.6                    .  .    .  .  21  30               0.82
```

Figure 8-15. A display of the factors and clocks used in the control algorithms at the reaction end.

The columns "LO LIMIT HI" are the low limits and high limits for the no-corrective action dead-band on the values of those measurements used as feedback signals to those loops on feedback control. As an example, provided the flash drum condenser vapor exit temperature remains between 48.0 and 52.0 degrees Centigrade, the cooling water will be on strict flow control to an unchanging target flow. However, if the temperature rises or falls outside these limits, then the target flow will be increased or decreased in some incremental fashion as prescribed in the control algorithm until the vapor exit temperature is returned to between these two limits.

Those values preceded by an "H" are history values. For the reactor level (H 77.7), this value is the current value because it is not on computer control. However, for the acidic crude product (H 65.4), it is the highest level to which the flash drum rose during its present excursion. By comparing the current level to the history level, the algorithm can tell whether the level is still rising or has begun to fall, and in consequence can choose between preprogrammed control alternatives. For Mares' Sweat, the pH went to 7.6 (H 7.6) before the current corrective action sequence took effect.

The Bubble:Double ratio and the Toil:Trouble ratio are operator entries. A prospective future development (generation two) of their respective control algorithms is to model the operator choice and close the loop on the ratios themselves. The long-term ratio (LTR) then becomes the ratio of choice should the loop mode be changed from feed-forward fixed ratio control to feedback variable ratio control. Provision of a long-term ratio will also allow "bumpless-transfer" when the mode is initially switched to feed-forward fixed ratio control from other modes. The long-term ratio provides secure storage for the historic ratio previously in use under feedback variable ratio control. It is not necessary for the operator to enter the desired value of the variable ratio when he switches modes over. The logic picks up the long-term ratio and provides a ramp feature to change the loop target gradually from the calculated target using the now discarded fixed ratio to that calculated using the long-term ratio. The LTR once again is subject to variation as mandated by the algorithm and the feedback parameter, while the loop mode remains at feedback variable ratio control.

If the Bubble load is changed, then the change is accommodated by a ramp to the new load value. The ramp is currently limited to one pulse every two measurement calculation decision cycles. The algorithm can detect a change in the load and will automatically create an initial value for an intermediate Bubble flow target in that cycle. Ordinarily, the initial value will be the actual mass flow when a load change is detected. This intermediate target will be incrementally altered within the prescribed ramp limitations in each successive cycle until it reaches the new load. It will be the target to which Bubble is flow controlled cycle to cycle while the ramp is in progress. At the new load, the ramp will be discontinued, and the Bubble flow target will be the Bubble load. All other loops under computer control are aware of a ramp in progress and may or may not choose intermediate targets as necessary.

To handle signal noise and to dampen control action, several excessive successive clocks for flow control have been implemented (ESC). In each of the algorithms where an excessive successive clock has been implemented different weightings have been assigned to various sizes of differences between actual flow and target flow. Only when the weighted running count exceeds the set-count limit is a control action predicated. The count is signed to give direction and to signal whether the actual flow is greater than or less than the target flow. On a change of position (greater/less), the count is zeroed. The purpose of the excessive successive counter is to filter out the minor flow differences, but still react to them if they are sustained over several cycles in one or another direction. However, the ESC must also give immediate corrective action to any large difference between actual flow and target flow. Bubble has a current running count of $+1$ with a limit of two. Reactor recycle has an ESC limit of 3, and reactor steam has an ESC limit of 10.

In Figure 8-14 Mares' Sweat was described as being in a wait period. In fact, the wait count is up to 21 out of a total wait of 30 counts. When the count exceeds 30, the wait will be over, and the Mares' Sweat algorithm will again use the pH signal to prescribe control action.

Some algorithms have been setup in such a fashion that the target flow is never to be set below a minimum flow or above a maximum flow. The target flow for the flash drum condenser cooling water must never be set at less than 40 gallons per minute (40 $<F$). The target flow for the flash drum cooler cooling water must never be set at less than 60 gallons per minute (60 $<F$).

Finally, the display has all the orifice plate factors in the respective flow units (OPF). Should transmitters be reranged, new orifice plates installed, or old orifice plates be recalibrated *in situ,* then the new factors can be entered by the operator; it is not necessary to patch or recompile.

This display gives the activities of the 13 loops on the computer at the purification end of the process. The display layout is very similar to that used for the reaction end. The 13 loops (in order) are: (1) extractor feed (EX-

```
CONTROL:  LOOP CONDITIONS                    PURIFICATION        1053    29 FEB 81

          MODE ADJ   GPM      #/h    CVP    TARGET    TVP       VARIABLE       DIV

EXFEED    C    1    25.90   4250.2   509   4240.7    510    1.541:1  WETRAT    7.51
FGSSPT    R    -2   39.91  17000.3   354  18000.1    352    4.245:1  DRYRAT    7.92

EX#1      C                          181              181      67.2   LEVEL%
EX#2      C                          401              401      65.3   LEVEL%
EX#3      C                          614              614      61.0   LEVEL%

EVSTM     L                 5130.4   180   5130.4    180      90.1    DEGC      7.23
EVBTM     A    -3    7.76   4250.5   529   4200.0    526      77.7    LEVEL%    7.64


DCFEED    U          7.57   4000.8   617   3999.0    617                        6.85
TSJ       C    1            2.789    213   2.800     214    1428:1              7.16
DCRFLX    C          94.29           362   94.31     362    32.5:1   EXTREFR    8.27
DC STM    C    6            4964.0   536   5000.0    542     128.4    DEGC      8.78

DC BTM    C    2     4.91   3210.2   321   3250.0    323      41.6    LEVEL%    6.89
DC CCW    C    -2   350              437   349       435      93.3    DEGC     10.00
```

Figure 8-16. A display of the control loop conditions at the purification end.

FEED), (2) Frogs' Spit (FGSSPT), (3) extractor drum #1 (EX#1), (4) extractor drum #2 (EX#2), (5) extractor drum #3 (EX#3), (6) evaporator steam (EVSTM), (7) evaporator bottoms (EVBTM), (8) distillation column feed (DCFEED), (9) tincture of Toad Stool Juice (TSJ), (10) distillation column reflux (DCRFLX), (11) distillation column steam (DC STM), (12) distillation column bottoms (DC BTM), and (13) distillation column condenser cooling water (DC CCW). The only features that will be described in detail are previously unexplained mode options and the variable column.

The extractor feed flow is the key flow for the purification end. Currently, the target flow rate is 4240.7 #/h. This rate is an operator-entered value. The Frogs' Spit is ratio controlled at a 4.245:1 solvent:solute dry basis (4.245:1 DRYRAT). For convenience, the volume:volume ratio is also calculated and displayed as the wet ratio (1.541:1 WETRAT). The dry-basis ratio is an operator-entered value. The Frogs' Spit flow is on computer feed-forward ratio control and is currently ramping (MODE R). The operator must have recently changed the dry ratio, since the Frogs' Spit target flow calculated as the product of the dry ratio and the extractor feed target flow is still being ramped toward its new value. The extractor drum levels are all on the computer under a level feedback algorithm. The evaporator steam is not on the computer (MODE L). The evaporator bottoms is on computer level feedback flow control with the actual level currently above the upper limit (MODE A).

The distillation column feed is on computer flow control and is being ramped up (MODE U). The ramp will terminate automatically when the distillation column feed target flow is equal to or greater than the extractor feed target flow. The tincture of TSJ is on computer feed-forward ratio control. The ratio is currently 1428:1. The distillation column reflux is on computer flow control. As an aid to the operator, the external reflux ratio is calculated and displayed (32.5:1 EXTREFR). The distillation column steam

```
CONTROL: FACTORS & CLOCKS              PURIFICATION       1052' 29 FEB 81

        LO LIMIT HI       LTR   RAMP    ESC     OOL    WAIT  MIN&MAX    OPF

EXFEED                           105    . .     . .     . .              3.45
FGSSPT                  4.200          -1  5    . .     . .              5.05

EX#1    25.0   75.0                     . .     . .     . .
EX#2    25.0   75.0                     . .     . .     . .
EX#3    25.0   75.0                     . .     . .     . .

EVSTM   90.5   91.5     0.300   101     . .     . .     . .  F< 5600   712.5
EVBTM   30.0   70.0             201     . .     . .     . .              1.02


DCFEED                           301    . .     . .     . .              1.11
  TSJ                   1500            . .     . .     . .              0.392
DCRFLX                  30.00           . .     . .     . .             11.49
DC STM 128.0  131.6     1.250           . .     . .     . .            570.6

DC BTM  35.0   65.0                     . .     . .     . .              0.72
DC CCW  93.1   95.9                     . .     . .     . .             35.
```

Figure 8-17. A display of the factors and clocks used in the control algorithms at the purification end.

is on computer temperature feedback flow control. The distillation column bottoms flow is on computer level feedback flow control. The distillation column condenser cooling water is on computer vent temperature feedback flow control. The flow transmitter is at its limit.

The variable values within each of the 13 control algorithms for the purification end of the process are displayed. The row titles and relative row positions correspond to Figure 8-16, and it is similar to Figure 8-15 in overall layout. The long-term ratios (LTR) are just reference values; variable ratio control has not been implemented on these loops yet. The ramps implemented to date are: (1) extractor feed, a very slow one pulse every five cycles (1@5); (2) evaporator steam, a slow one pulse every one cycle (1@1); (3) evaporator bottoms, two pulses every one cycle (2@1); and (4) distillation column feed, a fast three pulses every one cycle (3@1).

On Frogs' Spit, the excessive successive counter has a running -1 count on a limit of five. This is the first successive count that the actual flow has been below the target flow. For the previous cycle, it must have been above the target flow. No out-of-limit counters (OOL) have been implemented. These counters are used to slow down feedback action on feedback loops. The clock counters are setup so that a certain number of successive values of the feedback variable must be out of limit before the excursion can trigger any corrective action. As now configured on all the loops, the first value to exceed the limits initiates the corrective action. The evaporator steam has a maximum flow limit of 5600 pounds per hour ($F< 5600$).

These sample displays demonstrate the power, the flexibility, and the diversity of computer control. They also demonstrate the compactness of its human interface.

9

CONCLUSION—
A LOOK BEHIND AND AHEAD

The advance of microelectronic technology can be guaranteed. The production of the microelectronic industry is doubling every year, and it still remains a sellers' market. Production costs are dropping 10-20% annually in real terms because of the production learning curve. The lowered prices stimulate demand which in turn consumes the extra output. But beyond mass production's learning curve are the technological breakthroughs which are causing even steeper declines in costs. The real cost of processor memory declined 97% from 1968 to 1979!

There is also a rapid sophistication of available software. This is a powerful tool to the already trained, but it does raise the entry-level stakes. As much as sophisticated software demands powerful hardware, the improving hardware is fueling software sophistication. Peripherals are exploding in virtuosity and plummeting in price. Time-sharing is no longer an economic necessity. The move is toward dedicated systems in which the file editing, file handling, and languages can be customized to the convenience of the user. Indeed, tomorrow's computer will not be one powerful central processor surrounded by intelligent peripherals. The likelihood is that it will be a school of a score or more parallel processors sharing one common megabyte(s?) memory.

With the continued miniaturization of components, it is easy to predict a computer equal to today's most powerful which will fit comfortably inside a desk drawer and future process plant control rooms the size of offices. But this ignores two problem areas: the process front-end and the communication interface. As a computer-consuming industry, we must think in terms of 16 or 32 communication channels per installation, not one or two. We must think of the analog measurement interface in terms of 320 or 640 points, not 16 or 32. The incremental expansions must be in 64s or 96s, not twos and threes. We must pay much greater attention to termination panels and not talk of single-twisted pairs but 32-pair bundles, telephone-cable style.

Computer use is truly in its infancy, and we must not erect false barriers to imaginative implementation. We must not set forth a string of academic requirements as the main features of a job qualification. This is a brand new field. The self-taught and the experienced are still implementing. They are only just beginning to write books. They have had little time to teach. Indeed, universities and colleges are beginning only now to offer the necessary curricula. Much of the learning and many of the techniques are still "in-house" in industry. The problem is to wrest the established thinking from the mindless priesthood who have taught the machines only to address labels and add entries, who have hemmed us all in with an unthinking bureaucracy. We must change that established thinking into creating imaginative solutions by computer methods. Remember, two bicycle repairmen showed the world powered flight.

If these machines, these programmable processors, are to benefit us as individuals, then we as individuals must learn to use them. To ask anyone else to learn to use the machine on our behalf is to deny ourselves its very capacity to increase our productivity. But that moves us away from the topic of this book.

Today, there is no pool of trained people who can implement computer process control. An owner wishing to improve his process either must have an employee do it by trial and error under his own initiative or must try to hire the experience. The latter is a haphazard course because the employer is unsure about whom to hire and has no yardstick by which to judge the performance of the employee. Computer control is not an overnight miracle. One loop's logic can be written in five minutes, but it may take three weeks of real-time trial and error until the control is precise enough to be considered final. Perhaps a profession may emerge. It certainly would encourage career development. It definitely would improve the state of the art. Too much of the present practice that is poor is protected by the welfare state of all-purpose operating systems. It can be exposed only by the development and recognition of purpose-dedicated systems, purpose-dedicated to process improvement. The future will see an overall growth in expertise and knowledge. The depth of knowledge and the level of expertise is the responsibility of today's best practitioners in the art and science of process automation by programmable processors.

If you are still perturbed by the immensity of the software problem as you read the final pages of this book, be assured that many fine software packages are available either from the computer manufacturers, the instrument vendors, or even software houses. Nobody has to start from scratch. Total turnkey systems are available for both the development of software and the control of processes. There is a great variety of training available in all phases of implementation. Nothing at all stops you from buying a turnkey, "fill-in-the-blanks" system where the computer itself will prompt and query you as you construct a control algorithm, and then begin to allow trial use almost immediately. Perhaps that is all the application software sophistication you need for your purposes. But I believe that as you become familiar with your system and begin to realize the immediate benefits and the future potential, your own ingenuity and individual needs will at some point necessitate a certain customizing, a certain tailoring of your system. At that point, you may switch to writing your own algorithms in the higher languages, but this could be done on your current "fill-in-the-blanks" system without hardware modification. Or you may choose to install a completely new operating system and implement a mixture of fundamental control algorithms written at the assembly level, and economic optimization routines written in a higher language. Two conditions are certain: the computer market is so flexible and so dynamic that you will be able to find a system that meets your original needs and will grow as far and as fast as your own dictates require and you will find commercially available software that lets you do process control at your level of programming expertise.

From that point on, whether you grow and delve deeper or remain content is in your own hands.

A few closing thoughts:

1. A mass flow is smoothest and has the smallest standard deviation when the control valve is on manual.

2. The mass flow equation of $FLOW = OPF * \sqrt{GASFAC} * DIVV$ is as accurate as the methods of measurement.

3. A flow will respond to even a one-increment change in the position of the control valve. For some process-end parameters, this may be sufficient change to move them from one end of their scales to the other.

4. It is better to have 500 lines of simple, well-explained program and 500 lines of documentation so that an engineer can understand the program in an hour than it is to have 51 lines of ultratight code doing the same work but without documentation (so that it takes an engineer five days to work out and understand the program).

5. Most computer systems are set up to be convenient for system generation and programmers. They are not convenient from the owner's viewpoint. The programmer's viewpoint is that the system is one of many. The owner's viewpoint is that it is his *only* system. Consequently, for one it is general where it should be specific and, for the

other, it is specific where it should be general. The system should not need development to include extra analog inputs or communication channels. However, it should need development if the system is to be used for word processing or accounting.

6. Changes in any calculations should be permissible on-line.

7. All displays should be table-driven. It should require little training to prepare and install a new display.

8. All measurements taken should be instantaneous. Any smoothing required should be done to the calculated mass flows, not to the individual measurements. Setting limits on the maximum number of valve-change increments per calculation cycle infers an intimate knowledge of the process. If this is not the case, then the individual measurement that is expected to be the source of gross error should itself be the subject of limits and cross checks.

9. The analog controller exists to control the measured variable. Use the computer to adjust the set point. Do not use the computer to control the variable through the intermediary controller. If the computer must control the variable directly, then it must control the valve directly.

10. As a controlling device, the computer is failing in part of its capabilities if it cannot present the voltage input, digital value, and engineering value for each measured variable, or if it cannot present the value of any other calculated variable. All the factors used in the calculations must be patchable on line.

11. The computer must recognize and act upon any spike of any variable that is equal to or greater than a previously assigned magnitude. This means that all variables should be scanned often enough so that one of two consecutive scans sees the spike.

12. The machine must have the capability of data collection. Such capability must be flexible and very easy to use. Without it, progress in control of the process is not possible, and knowledge of the process cannot be furthered.

13. For any control loop attempted, there must be several choices of both control logic and the final process-end variable monitored. The final process-end variable is that element whose state is used to close the control loop on "closed loop" control.

14. For continued forward progress, all subroutines in the machine must be documented to the extent that people initially unfamiliar with the purpose and use of them can on their own write subroutines using code that will work correctly from the first use. It is at these fun-

damental building-block levels that the documentation is most needed.

15. To display misleading levels of significance, unusual physical units, and unusual formats is to cause confusion. Each and every output should be understood by all who come in contact with the machine. Furthermore, changes in numbers displayed on CRTs should occur only when the change is significant. Avoid degrees Kelvin and pounds square inch absolute. Round off mass flows to no more than the fourth significant figure. Never imply an accuracy greater than you really have with displayed numbers. Never use a decimal when an integer will do. Never use an E-format if you possibly can avoid it.

APPENDIX

This appendix contains the following tables:

1. The International System of Units

2. Conversion factors:
   a. English into metric units
   b. Metric into English units

3. Capacity of horizontal cylindrical tanks

THE INTERNATIONAL SYSTEM OF UNITS (SI)

	Quantity	INTERNATIONAL SYSTEM SI or MKSA			SUBMULTIPLE SYSTEM CGS			OTHER MULTIPLES AND SUBMULTIPLES OF SI
		Dimension	Unit	Symbol	Unit	Symbol	SI equivalent	
BASE UNITS	Length	L	metre	m	centimetre	cm	10^{-2} m	10^3 m = 1 km : 10^{-3} m = 1 mm : 10^6 m = 1 micron (μ) : 10^{-10} m = 1 angström (Å)
	Mass	M	kilogramme	kg	gramme	g	10^{-3} kg	10^3 kg = 1 tonne (t) ; 10^2 kg = 1 quintal (q)
	Time	T	second	s	second	s	s	3 600 s = 1 heure (h) : 60 s = 1 minute (min) ; 10^{-3} s = 1 millisecond (ms)
	Electric current		ampere	A				10^{-3} A = 1 milliampère (mA) : 10^{-6} A = 1 microampère (1μA)
	Thermodynamic Temperature		kelvin	K				Temperature Celsius (°C) is the difference between two thermodynamic temperatures $T-T_0$ where T_0 = 273,15 kelvins
	Amount of substance		mole	mol				
	Luminous intensity		candela	cd				
SUPPLE-MENTARY	Angle, plane Angle, solid		radian stéradian	rad sr				
DERIVED UNITS	Area	L^2	square metre	m²	square centimetre	cm²	10^{-4} m²	10^4 m² = 1 hectare (1 ha) : 10^2 m² = 1 are (a) : 10^{-6} m² = 1 mm²
	Volume	L^3	cubic metre	m³	cubic centimetre	cm³	10^{-6} m³	10^{-3} m³ = 1 dm³ = 1 litre (l)
	Force	MLT^{-2}	newton	N	dyne	dyn	10^{-5} N	10^4 N = 10^3daN : 10^3N = 1 kilonewton (kN) : 10N = 1 décanewton (daN)
	Energy - Work Heat	ML^2T^{-2}	joule	J	erg	erg	10^{-7} J	10^6J = 1 kilonewton·km (kN·km) : 10J = 1 daN · m : 1J = 1 Nm : 4,18J = 1 calorie (cal) : 4,18 10^3J = 1 thermie (th)
	Power	ML^2T^{-3}	watt	W	erg par seconde	erg/s	10^{-7} W	10^6W = 1 mégawatt (1 MW)
	Velocity linear	LT^{-1}	metre per second	m/s	centimetre per second	cm/s	10^{-2} m/s	$\frac{1}{3,6}$ m/s = 1 km/h ; $\frac{1 852}{3 600}$ m/s = 1 knot
	Acceleration	LT^{-2}	metre per second squared	m/s²	gal	cm/s²	10^{-2} m/s²	
	Pressure	$ML^{-1}T^{-2}$	pascal	Pa	barye	dyn/cm²	10^{-1} Pa	10^2 Pa = 1 hectobar (hbar) ; 10^5 Pa = 1 bar : 10^2 Pa = 1 millibar (mbar)
	Viscosity, dynamic	$ML^{-1}T^{-1}$	pascal-second	Pa.s	poise	P	10^{-1} Pa. s	10^{-3}Pa. s = 1 centipoise (cP)
	Viscosity kinematic	L^2T^{-1}	metre squared per second	m²/s	stokes	St	10^{-4} m²/s	10^{-6} m²/s = 1 centistoke (cSt)

CONVERSION FACTORS
ENGLISH INTO METRIC UNITS

to obtain ✓		multiply the number of ✓		by ✓
0.0393701	Millimètres	Inches (pouces)	in	25.4
3.28084	Mètres	Feet (pieds)	ft	0.3048
1.09361	Mètres	Yards	yd	0.9144
0.621373	Kilomètres	Statute miles (milles terrestres)		1.60934
0.539613	Kilomètres	**Nautical miles (UK)**		
		(milles marins anglais)		1.85318
0.539957	Kilomètres	**Nautical miles**		
		(milles marins-autres pays)		1.852
0.155	Centimètres carrés	Square inches (pouces carrés)	in², sq in.	6.4516
10.7639	Mètres carrés	Square feet (pieds carrés)	ft², sq ft	0.0929
2.47105	Hectares	Acres		0.404686
0.386102	Kilomètres carrés	Square miles (milles carrés)	sq mile	2.58999
0.0610236	Centimètres cubes	Cubic inches (pouces cubes)	in³, cu in.	16.3871
0.0353147	Décimètres cubes	Cubic feet (pieds cubes)	ft³, cu ft	28.3168
0.264178	Décimètres cubes	Gallons (US)	gal (US)	3.78533
0.219976	Décimètres cubes	Gallons (UK)	gal (UK)	4.54596
35.3147	Mètres cubes	Cubic feet	ft³, cu ft	0.0283168
6.28994	Mètres cubes	Barrels (US) (barils)	bbl	0.158984
150.959	Mètres cubes par heure	Barrels per day (barils par jour)	bbl/day	0.00662433
15.4324	Grammes-force	Grains-force	grf	0.0647989
0.035274	Grammes-force	Ounces-force (onces-force)	ozf	28.3495
2.20462	Kilogrammes-force	Pounds-force (livres-force)	lbf	0.453592
0.224809	Newtons	Pounds-force	lbf	4.44822
0.0234534	Kilogrammes-force	Sacks (cement)		42.6377
1.10231	Tonnes-force	Short tons-force (t -force USA)	sh tonf	0.907185
0.984204	Tonnes-force	Long tons-force		1.01605
0.671971	Kilogrammes-force	Pounds-force		
	par mètre	per foot	lbf/ft	1.48816
8.34523	Kilogrammes-force	Pounds-force		
	par décimètre cube	per gallon (US)	lbf/gal	0.119829
62.4278	Kilogrammes-force	Pounds-force		
	par décimètre cube	per cubic foot	lbf/ft³	0.0160185
0.3505	Kilogrammes-force	Pounds-force		
	par mètre cube	per barrel	lbf/bbl	2.85307
0.0805214	Litres par mètre	Gallons (US) per foot	gal/ft	12.4191
14.5038	Bars	Pounds-force per square inch	lbf/in², psi	0.0689476
14.2233	Kilogrammes-force	Pounds-force		
	par centimètre carré	per square inch	lbf/in², psi	0.070307
0.711167	Kilogrammes-force	Short tons-force		
	par millimètre carré	per square inch	tonf/in²	1.40614
102.408	Kilogrammes-force	Short tons-force		
	par millimètre carré	per square foot	tonf ft²	0.00976486
0.737561	Joules	Feet-pounds-force	ft lbf	1.35582
7.23301	Kilogrammètres	Feet-pounds-force	ft lbf	0.138255
0.737562	Mètres-newtons	Feet-pounds-force	ft lbf	1.35582
0.684944	Tonnes-force-kilomètres	Short tons-force-miles		1.45997
0.00134102	Watts	Horse powers	hp	745.7
0.98632	Chevaux-vapeur	Horse powers	hp	1.01387
0.000947813	Joules	British thermal units	Btu	1055.06
3.96707	Kilocalories	British thermal units	Btu	0.252075
0.368553	Kilocalories	British thermal units		
	par mètre carré	per square foot	Btu/ft²	2.71331
1.79943	Kilocalories	British thermal units		
	par kilogramme	per pound	Btu/lb	0.55573
0.112335	Kilocalories	British thermal units		
	par mètre cubé	per cubic foot	Btu/ft³	8.90196
$°C \cdot \frac{9}{5} \cdot 32$	Degrés Celcius	Degrees Fahrenheit	°F	$(°F \cdot 32) \frac{5}{9}$
5.61448	Mètre cube	Cubic feet		
	par mètre cube	per barrel (US)	ft³/bbl	0.178111
0.042	Litres par mètre cube	Gallons (US) per barrel (US)	gal/bbl	23.8095

by ↖	↖ multiply the number of	↖ to obtain		

CONVERSION FACTORS
METRIC INTO ENGLISH UNITS

CAPACITY OF HORIZONTAL CYLINDRICAL TANKS
Mensuration of tank : capacity = V, height = H

Fraction of H	Fraction of V	Fraction of H	Fraction of V	Fraction of H	Fraction of V
0.01	0.0017	0.34	0.2998	0.67	0.7122
0.02	0.0047	0.35	0.3119	0.68	0.7241
0.03	0.0087	0.36	0.3241	0.69	0.7360
0.04	0.0134	0.37	0.3364	0.70	0.7477
0.05	0.0187	0.38	0.3487	0.71	0.7593
0.06	0.0245	0.39	0.3611	0.72	0.7708
0.07	0.0308	0.40	0.3736	0.73	0.7821
0.08	0.0375	0.41	0.3860	0.74	0.7934
0.09	0.0446	0.42	0.3986	0.75	0.8045
0.10	0.0520	0.43	0.4111	0.76	0.8155
0.11	0.0599	0.44	0.4237	0.77	0.8263
0.12	0.0680	0.45	0.4364	0.78	0.8369
0.13	0.0764	0.46	0.4490	0.79	0.8474
0.14	0.0851	0.47	0.4617	0.80	0.8576
0.15	0.0941	0.48	0.4745	0.81	0.8677
0.16	0.1033	0.49	0.4872	0.82	0.8776
0.17	0.1127	0.50	0.5000	0.83	0.8873
0.18	0.1223	0.51	0.5128	0.84	0.8967
0.19	0.1323	0.52	0.5255	0.85	0.9059
0.20	0.1424	0.53	0.5383	0.86	0.9149
0.21	0.1526	0.54	0.5510	0.87	0.9236
0.22	0.1631	0.55	0.5636	0.88	0.9320
0.23	0.1737	0.56	0.5763	0.89	0.9401
0.24	0.1845	0.57	0.5889	0.90	0.9480
0.25	0.1955	0.58	0.6014	0.91	0.9554
0.26	0.2066	0.59	0.6140	0.92	0.9625
0.27	0.2179	0.60	0.6264	0.93	0.9692
0.28	0.2292	0.61	0.6389	0.94	0.9755
0.29	0.2407	0.62	0.6513	0.95	0.9813
0.30	0.2523	0.63	0.6636	0.96	0.9866
0.31	0.2640	0.64	0.6759	0.97	0.9913
0.32	0.2759	0.65	0.6881	0.98	0.9952
0.33	0.2878	0.66	0.7002	0.99	0.9983

Example : Let us take a tank with capacity V = 12 000 l and height H = 2 m. Measurements show a liquid height of 0,20 m in the tank. How much liquid does the tank contain.
Answer : Fraction of height 0,20/2 = 0,10, corresponding to a volume fraction of 0,0520 in the table. The contenance is thus : 0,0520 × 12 000 = 624 l.

SUGGESTED READING

Instrumentation, Flow of Fluids, Conventional Control

Applied Instrumentation in the Process Industries (Gulf Pub. Co.)
Volume 1/2nd ed.: A Survey
   W.G. Andrews and H.B. Williams
Volume 2/2nd ed.: Practical Guidelines
   W.G. Andrews and H.B. Williams
Volume 3/2nd ed.: Engineering Data and Resource Material
   W.G. Andrews and H.B. Williams
Volume 4: Control Systems: Theory, Troubleshooting, and Design
   L.M. Zoss
Volume 5: Instrument Engineering Programs for Handheld Calculators
(Magnetic program cards also available.)
   Stanley W. Thrift
Control Valve Handbook, Fisher Controls Co. 2nd edition, 1977.
Flow of Fluids Through Valves, Fittings and Pipe, Crane Tech.
Paper #410, Chicago, 1969.
Handbook of Valves, Piping and Pipelines, Gulf Publishing Co., Houston, TX, 1982.
Equivalent Valves Reference Manual/18th Edition, Gulf Publishing Co., Houston, TX, 1982.
Anderson, N.A., *Instrumentation for Process Measurement and Control,* Chilton Book Co., Radnor, PA.
Kane, L.A., *Process Control and Optimization Handbook for the HPI,* Gulf Publishing Co., Houston, TX, 1980.
Shinskey, F.G., *Process Control Systems,* McGraw-Hill Book Co., New York, 1967.
Spink, L.K., *Principles and Practice of Flow Meter Engineering,* The Foxboro Co., Foxboro, MA.
Zappe, R.W., *Valve Selection Handbook,* Gulf Publishing Co., Houston, TX, 1981.

Computer Control

Smith, C.L., *Digital Computer Process Control,* Intnatl. Textbook, Scranton, PA, 1972.

Magazines

Hydrocarbon Processing, Gulf Publishing Co., Houston, TX
Byte, Byte Publications, Peterborough, NH
Kilobaud Microcomputing, Wayne Green, Inc., Peterborough, NH

Computer Hardware and Software

Abrams, M.D. and Stein, P.G., *Computer Hardware and Software: An Interdisciplinary Introduction,* A-W, 1973.
Bell, C.G., Mudge, J.C., and McNamara, J.E., *Computer Engineering,* Digital Press, Bedford, MA, 1978.
Eckhouse, R.H., *Minicomputer Systems: Organization and Programming,* Prentice-Hall, Inc., Englewood Cliffs, NJ, 1975.
Korn, G.A., *Minicomputers for Engineers and Scientists,* McGraw-Hill Book Co., New York, NY, 1973.
Maurer, W.D., *Programming; An Introduction to Computer Techniques.* rev. 2nd ed., Holden Day, 1972.
McNamara, J.E., *Technical Aspects of Data Communication,*
Digital Press, Bedford, MA, 1978.
Murphy, J.S., *Basics of Digital Computers, Volumes 1-3,* Hayden, 1970.

Digital Equipment Corporation Handbooks

Microcomputer & Memories Handbook
Microcomputer Interface Handbook
PDP-11 Peripheral Handbook
PDP-11 Programming Fundamentals Handbook
PDP-11 Software Handbook
PDP-11 Processor Handbook
PDP-8/A Minicomputer Handbook
PDP-8/E/F/M Small Computer Handbook
Terminals & Communications Handbook

Monograph Series

Mellichamp, D.A., editor, CACHE Monograph Series in Real-Time Computing, CACHE, Cambridge, MA, 1977.

Courses

Computer Process Control, Sponsor: AIChE Today Series, Several times a year at various locations.
Modern Digital Process Control, Sponsor: McGraw-Hill Seminar Center, Several times a year at various locations.
Digital Computer Process Control, Sponsor: Univ. of Colorado, Dept. of Chemical Engineering, Twice yearly.

INDEX